IMAGES
of America

DETROIT'S HISTORIC WATER WORKS PARK

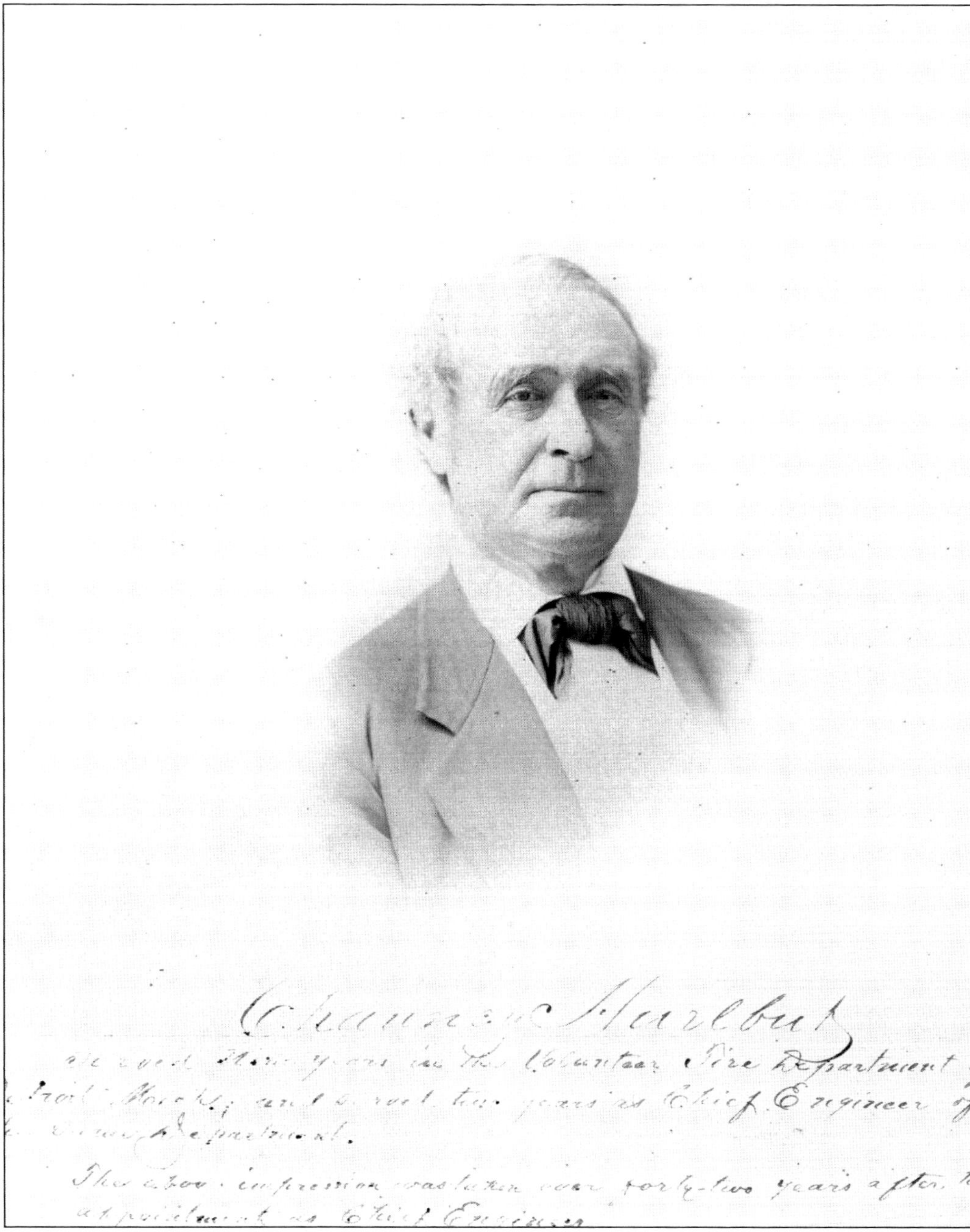

Chauncy Hurlbut earned his fortune in the grocery trade after coming to Detroit from New York in 1825. He was president of the Detroit Board of Water Commissioners when it purchased Water Works Park in 1874. He bequeathed the bulk of his $250,000 estate (equal to more than $6 million in 2011) to the City of Detroit for the beautification of Water Works Park. (Burton Collection.)

On the Cover: The Standpipe water tower and original pumping station at Water Works Park are pictured about 1890. The water tower was among the tallest of the world's structures for many years. (Detroit Water and Sewerage Department.)

IMAGES
of America

Detroit's Historic Water Works Park

Michael Daisy

ARCADIA
PUBLISHING

ISBN 978-0-7385-9363-0

Published by Arcadia Publishing
Charleston, South Carolina

Printed in the United States of America

Library of Congress Control Number: 2012934539

For all general information, please contact Arcadia Publishing:
Telephone 843-853-2070
Fax 843-853-0044
E-mail sales@arcadiapublishing.com
For customer service and orders:
Toll-Free 1-888-313-2665

Visit us on the Internet at www.arcadiapublishing.com

This book is dedicated to Beverly Daisy (1944–2003). I am further indebted to my wife, Cathy Daisy, for her love and support.

Contents

ACKNOWLEDGMENTS

Those who provided invaluable assistance in this project include Mark Bowden of the Burton Collection, Detroit Public Library; David Brody, P.E.; Gregory Smith; S.R. Boland; James Heath; Paul Gomez; and Irving Schuraytz, P.E.

Photograph sources are credited with their captions. Some recurring sources are abbreviated as follows:

(Author) Author's personal collection;
(Burton Collection) Burton Collection, Detroit Public Library;
(*Detroit News*) *The Detroit News*;
(DWSD) Detroit Water & Sewerage Department;
(DPC/LOC) Detroit Publishing Company/Library of Congress.

INTRODUCTION

Detroit's historic Water Works Park has been a continuous source of drinking water for Detroit and surrounding communities since 1879. That year, it became home to the city's first water treatment plant—for a time, the world's largest. In 2003, Detroit's newest water treatment plant debuted at the park site. The park, however, was always intended to be more than just a collection of pumps and pipes to distribute water from the Detroit River.

Located across the American channel of the Detroit River from Belle Isle, the largest city-owned island park in the United States, Water Works Park was intended to serve as a defining symbol of a proud city on the edge of the great American frontier. It was already world-famous and one of the city's most popular attractions long before Detroit was known as the "Motor City." Water Works Park harkens back to a time when Detroit, with its wide tree-lined boulevards, was known to the world as the "Paris of the Midwest."

The site on which the park sits was once a part of what French settlers knew as the Grand Marais, or "Big Marsh." The marsh was a strip of land covered with grasses and wildflowers that extended along the Detroit River and Lake St. Clair roughly from the park's current 10100 East Jefferson Avenue location to around Bishop Road in neighboring Grosse Pointe. Near the northern end is a large point, or *grosse pointe*, extending into Lake St. Clair and known as Windmill Pointe.

The point is named for an 18th-century French windmill that was later succeeded by one of the earliest lighthouses on the upper Great Lakes. Windmill Pointe and Belle Isle were both considered as locations for water-pumping stations prior to the selection of Water Works Park. Windmill Pointe was rejected because of its relative costs and distance; Belle Isle, because of its isolation. Access to the big island was sporadic and unreliable prior to the construction of the bridge connecting it to the mainland in the 1880s.

Annual autumn rains routinely submerged land in the Grand Marais. Frigid winter temperatures converted the marsh into a sea of glass, making it a popular site for wintertime recreation during the region's French occupation during the first half of the 18th century. Young people gathered there to dance, feast, and race their carioles (cutters or sleighs) during daylight hours. At the sound of the evening gun, the revelers returned to the safety of Fort Pontchartrain—which became known as Fort Detroit around 1750—for protection from potential nighttime depredations by the indigenous Indian population.

Led by its president, Chauncy Hurlbut, the Detroit Board of Water Commissioners paid Robert P. Tom $35,000 in January 1874 for a 56-acre plot of land he owned below Connor's Creek in Hamtramk Township, between East Jefferson Avenue and the Detroit River. Tom's Grand Marais Farm soon became known as Water Works Park. Additional land purchases between 1906 and 1912 increased the size of the park to its present 110 acres.

At its operational birth in 1879, the park was a technological marvel and the culmination of earlier efforts to establish a water distribution system in 1824 with a single pump, on a wharf at the foot of Randolph Street. Though it has been a reliable source of high-quality drinking water for more than 130 years, longevity is not the park's only claim to fame. For many years, it also functioned as a popular public gathering space.

Water Works Park was a popular tourist destination decades before December 1923, when it became the site of the world's largest water treatment plant. Visitors came to see the park's world-famous structures and intricate floral displays, and to immerse themselves in the serenity of the grounds. People came from great distances, and no trip to Detroit was complete without a visit to Water Works Park. By 1893, the park contained a man-made winding canal with two islands, three bridges, and a wading pool. The canal accommodated rowboats and other small craft entering from the Detroit River. Anglers were rewarded with swarms of fish, attracted to the warmer waters of the park's coaling canal and watercourse.

In the "Report of the Superintendent of Grounds" contained within the 1894 *Annual Report of the Detroit Board of Water Commissioners*, Elbridge A. "Scrib" Scribner—the first superintendent of grounds at Water Works Park—states, "Facts and figures show [Water Works Park] to be a pumping station that for efficiency and beauty is equaled by few and excelled by none . . . The surrounding Park is rapidly coming into a high state of cultivation, and the multitudes who visit it testify that as a place of rest and recreation it is one of which the citizens of Detroit may justly feel proud."

In a fit of civic pride in 1910, members of Detroit's common council determined that Water Works Park was too ordinary a name. A proper name—one that conveyed the appropriate ambiance and personality of such a magnificent space—was needed. Council members chose Gladwin Park, for Maj. Henry Gladwin, the dashing commander of Fort Detroit in 1763, who defended the tiny outpost against a siege led by Chief Pontiac.

Major Gladwin was certainly a gallant enough figure from Detroit's distant past, but Detroiters refused to accept the name. It appears on a few documents, and continues to surface now and then in conjunction with these hallowed grounds, but Detroiters continue to refer to the site as Water Works Park, a perfectly utilitarian name that accurately describes the park's primary function and reason for being.

One

The Grand Marais

Prior to the establishment of the new works at Water Works Park, Detroit's development into an important manufacturing center after the Civil War spurred enormous population growth. The old Orleans Street works was setting new records for the volume of water pumped, but was hard-pressed to keep pace with growing demand from manufacturing and residential usage. Furthermore, the works was closed in on all sides by industries and faced "an extreme fire hazard."

At the time of its purchase in January 1874, Water Works Park was known as the Grand Marais Farm, or "the farm at the big marsh." It was so called because about 10 acres of it were submerged in shallow water.

Detroit mayor Hugh Moffat favored expanding the Orleans Street Works because he believed it made more financial sense. He opposed the Hamtramck site because it was swampy, and Moffat thought any water supply originating there would be unhealthy. He also considered the Detroit Board of Water Commissioners too independent for their own or the public's good, and he opposed their selection on principle.

Led by former water commissioner John Ruehle, the city council approved the sale of bonds to finance construction of a new waterworks in Hamtramck, but it was vetoed. After much bickering, council members passed another resolution by unanimous vote, which was also vetoed. The measure was eventually passed by a unanimous vote to overturn the veto, and the way was cleared for construction of a new facility on the grounds that became Water Works Park. But the dispute left doubts in the public mind about the quality of water being pumped from the new site.

A "Hydraulic Commission," comprised of high-ranking engineers in the Army Corps of Engineers, inspected the site in the spring of 1874. Their report supported the board's selection. Board members, however, hedged their bets.

To alleviate concerns about water quality, reports stressed that water from the new location was of "unquestionable quality." Detroit's public health officer, Dr. Elisha Leach, also gave assurances that "water from the marsh and from Connor's Creek could not reach the mouth of the inlet pipe."

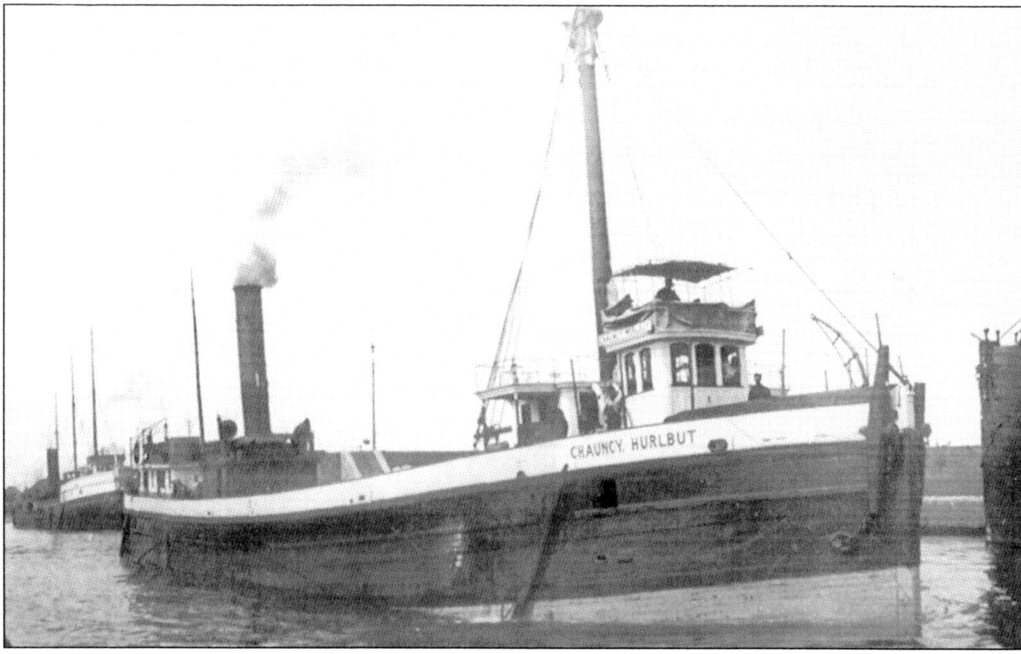

The *Chauncy Hurlbut* was a propeller-driven steam barge built by the Chandler Brothers of Detroit and launched on Thanksgiving Day, 1873. The little vessel traversed the waters of the Great Lakes for more than 30 years. She was sunk near Whitefish Point in Lake Superior in 1909 after being badly battered in a storm. (DWSD.)

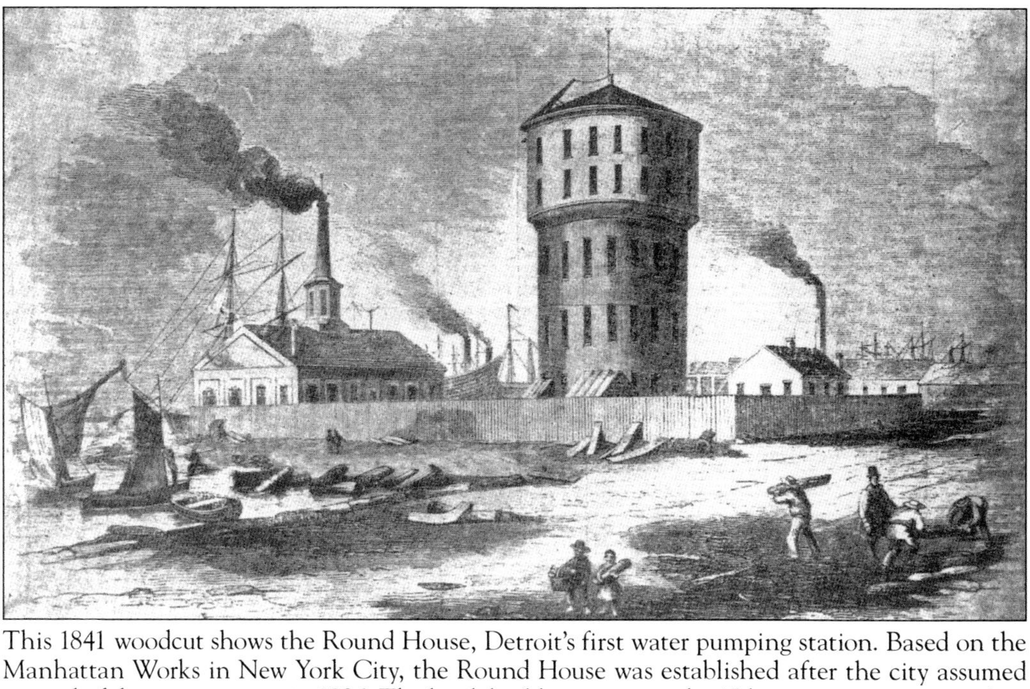

This 1841 woodcut shows the Round House, Detroit's first water pumping station. Based on the Manhattan Works in New York City, the Round House was established after the city assumed control of the water system in 1836. The brick building contained a 45-horsepower engine and an iron reservoir with a capacity of 422,979 gallons. In 1850, it was joined by the works established at the foot of Orleans Street. (DWSD.)

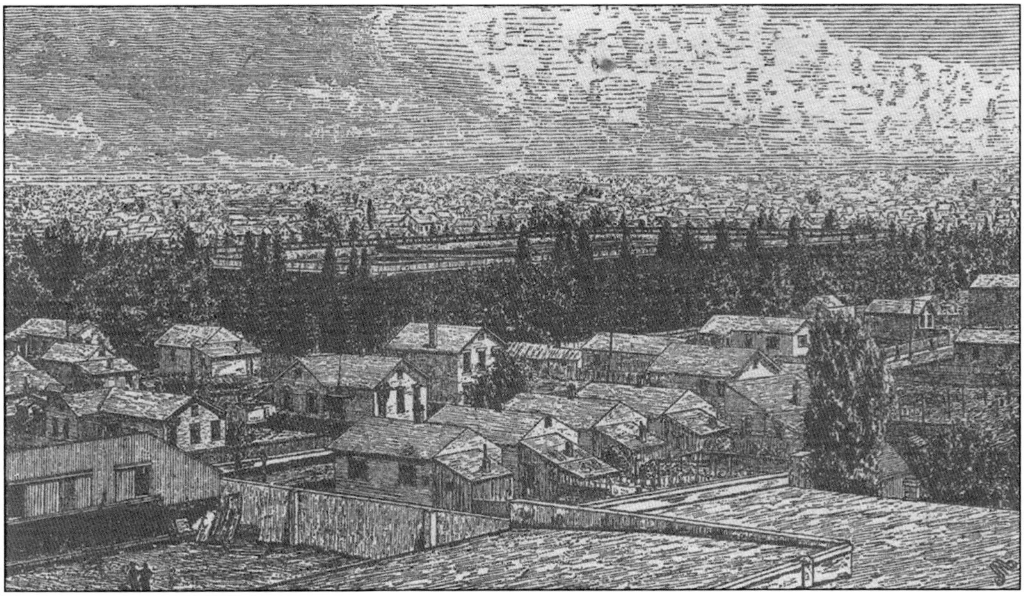

The reservoir depicted in this 1853 woodcut was established on four acres of land that was formerly a part of the Mullet Farm, located on the "extreme outskirts" of the city. The site on which the reservoir was located is in the current vicinity of Eastern Market on Detroit's near east side. (DWSD.)

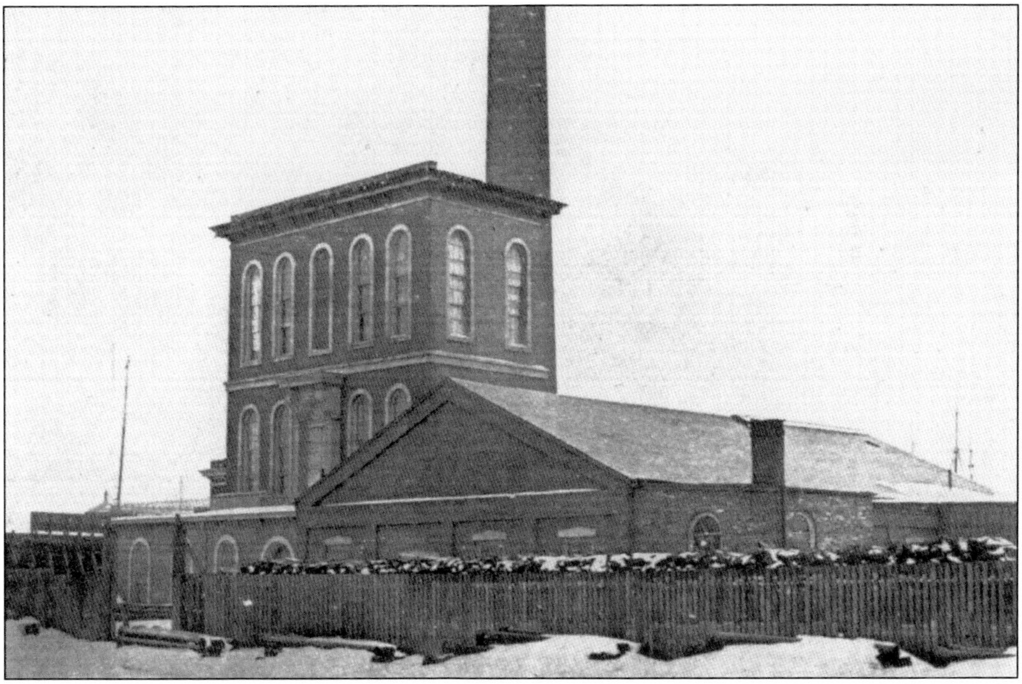

Placed in service in 1841, the Orleans Street Works, located near the present site of Chene Park, struggled to supply the demands of Detroit's rapidly growing residential and industrial population at the beginning of the 1870s. Unfortunately, expansion of the station was not possible. The crush and concentration of industry in its immediate vicinity was such that it "now faced an extreme fire hazard." (DWSD.)

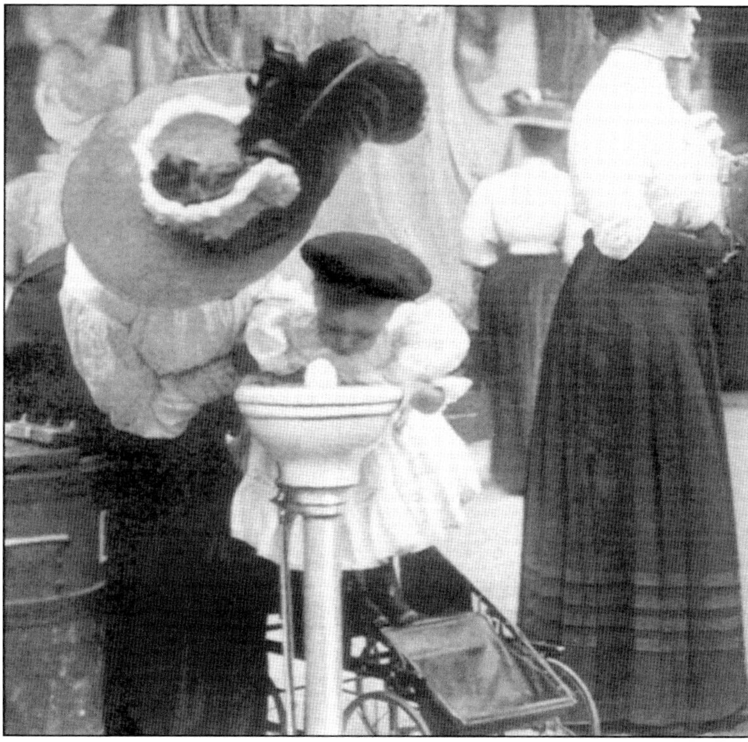

A young girl is assisted by her mother at a drinking fountain in this photograph from around 1900. The Detroit Board of Water Commissioners spent $255.24 to install public drinking fountains in 1871. Many of those initial fountains were combination drinking fountains, capable of providing service to both "man and beast." A few of them even included provisions for dogs. (DWSD.)

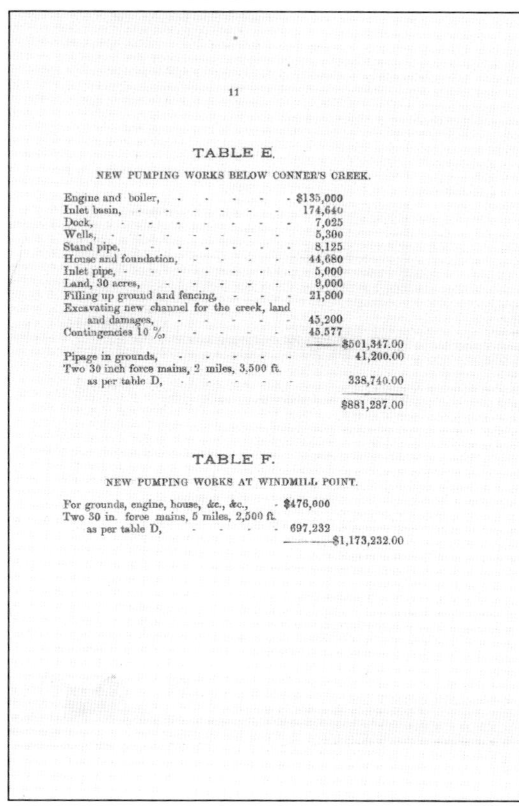

11

TABLE E.

NEW PUMPING WORKS BELOW CONNER'S CREEK.

Engine and boiler,	$135,000	
Inlet basin,	174,640	
Dock,	7,025	
Wells,	5,300	
Stand pipe,	8,125	
House and foundation,	44,680	
Inlet pipe,	5,000	
Land, 30 acres,	9,000	
Filling up ground and fencing,	21,800	
Excavating new channel for the creek, land and damages,	45,200	
Contingencies 10 %,	45.577	
		$501,347.00
Pipage in grounds,		41,200.00
Two 30 inch force mains, 2 miles, 3,500 ft. as per table D,		338,740.00
		$881,287.00

TABLE F.

NEW PUMPING WORKS AT WINDMILL POINT.

For grounds, engine, house, &c., &c.,	$476,000	
Two 30 in. force mains, 5 miles, 2,500 ft. as per table D,	697,232	
		$1,173,232.00

In 1872, the Detroit Board of Water Commissioners hired David Farrand Henry specifically to oversee the construction of a new waterworks. His original December 1872 estimate compared the potential costs of building a waterworks below Conner's Creek (today's Water Works Park) versus a new works at Windmill Pointe on Lake St. Clair. (DWSD.)

D.F. Henry's original 1872 estimate also compared the costs of establishing a waterworks on Belle Isle with expanding the existing works on the Detroit River at the foot of Orleans Street. Expansion of the Orleans works was rejected due to lack of space, as it was in a heavily congested and industrialized area. (DWSD.)

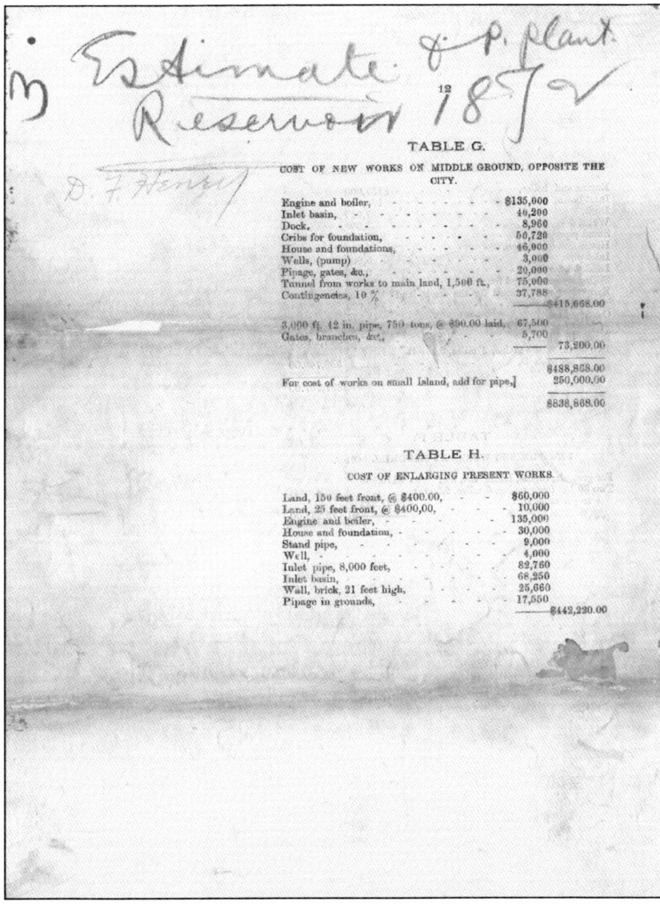

Estimate & P. plant
Reservoir 1872
D. F. Henry

12

TABLE G.

COST OF NEW WORKS ON MIDDLE GROUND, OPPOSITE THE CITY.

Engine and boiler,	$135,000	
Inlet basin,	40,200	
Dock,	8,960	
Cribs for foundation,	50,720	
House and foundations,	46,000	
Wells, (pump)	3,000	
Pipage, gates, &c.,	20,000	
Tunnel from works to main land, 1,500 ft.,	75,000	
Contingencies, 10 %	37,788	
		$415,668.00
3,000 ft. 12 in. pipe, 750 tons, @ $90.00 laid,	67,500	
Gates, branches, &c.,	5,700	
		73,200.00
		$488,868.00
For cost of works on small Island, add for pipe,]		250,000.00
		$838,868.00

TABLE H.

COST OF ENLARGING PRESENT WORKS.

Land, 150 feet front, @ $400.00,	$60,000	
Land, 25 feet front, @ $400,00,	10,000	
Engine and boiler,	135,000	
House and foundation,	30,000	
Stand pipe,	9,000	
Well,	4,000	
Inlet pipe, 8,000 feet,	82,760	
Inlet basin,	68,250	
Wall, brick, 21 feet high,	25,660	
Pipage in grounds,	17,550	
		$442,220.00

A portion of a 1925 map shows the three sites considered for Detroit's new pumping station in 1872: Water Works Park (Gladwin Park), Belle Isle, and Windmill Pointe, at the far right. Windmill Pointe was rejected because of distance and cost, while Belle Isle, since there was no bridge connecting it to the mainland at the time, was rejected for its relative inaccessibility. (David Brody, P.E.)

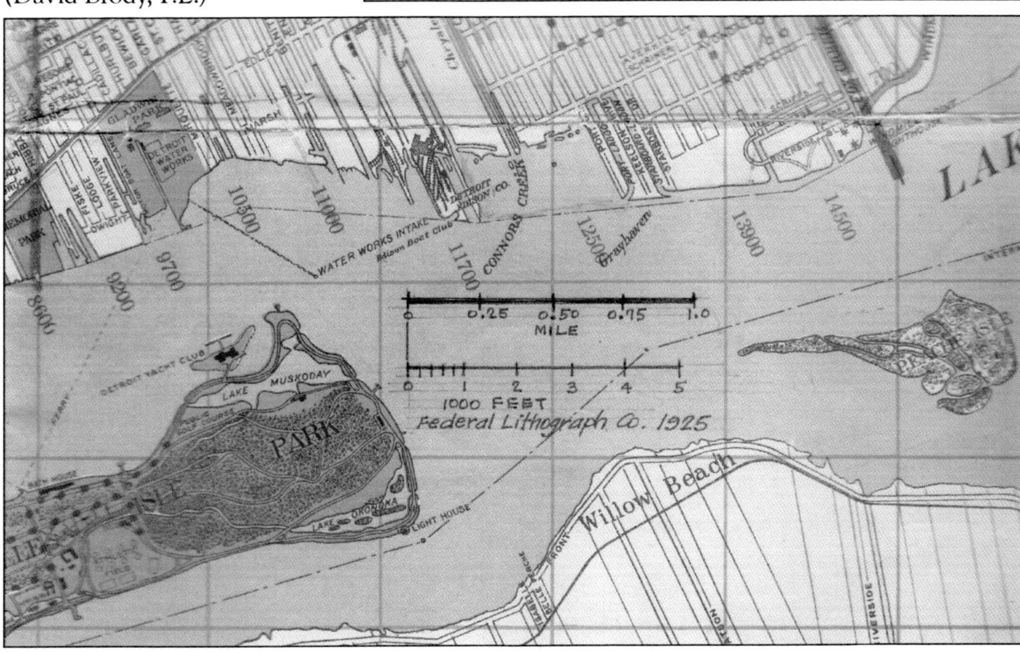

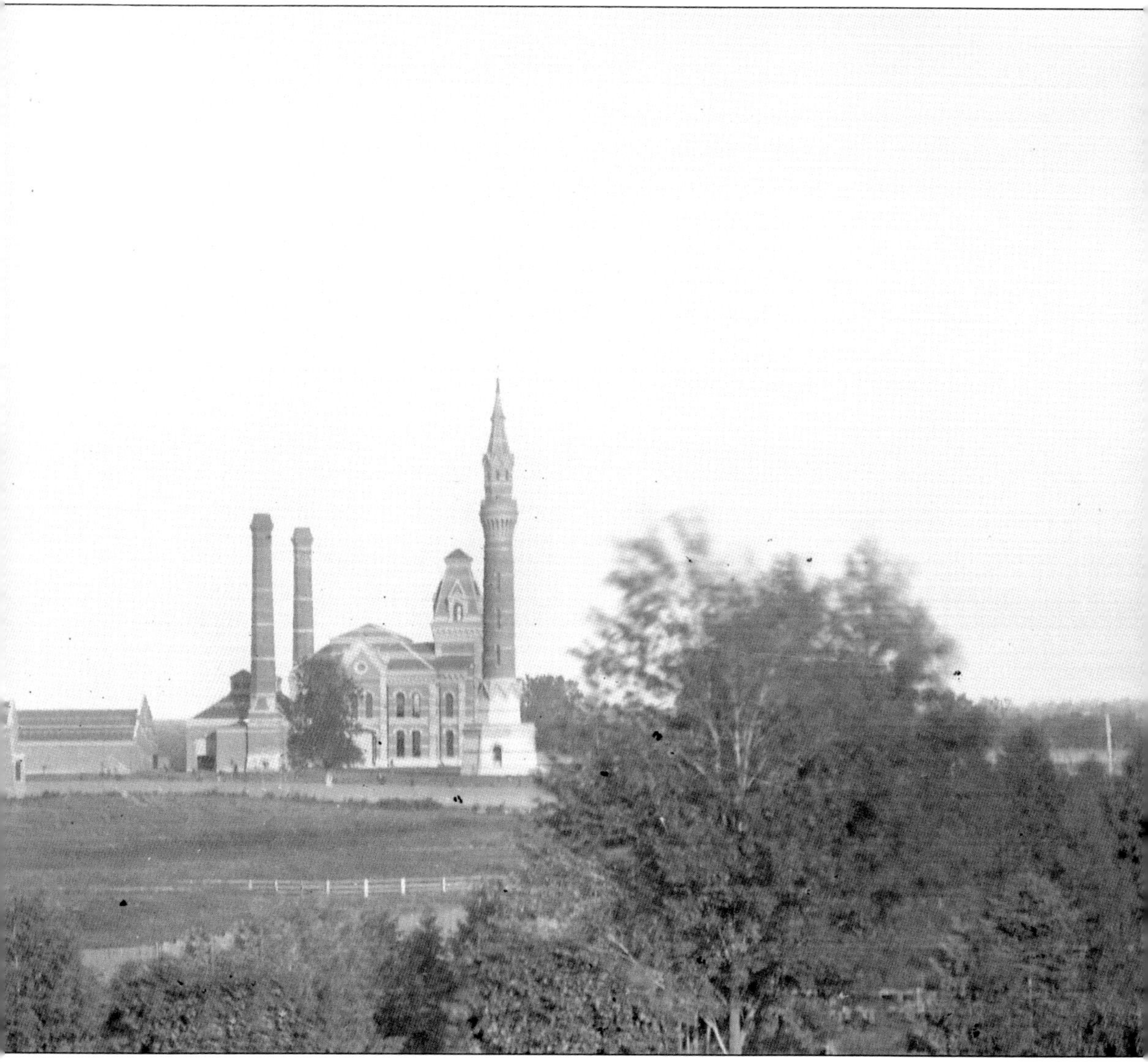

A distant view shows the original pumping station and water tower at Water Works Park in the early 1880s. The white fence suggests that the park might not have fully taken on the role of water supply utility yet. The tree in front of the station is undoubtedly one of the "Twelve Apostles," missionary pear trees named for the apostles that French settlers planted more than a century before. (Burton Collection.)

Two

Popular Urban Space

The primary purpose of Water Works Park has always been as a source of high-quality drinking water for Detroit and its surrounding regions. But in the 19th century, Detroit and other heavily industrialized American cities were crowded, dirty, noisy places, so city planners routinely provided open spaces where residents could escape from the hustle and bustle of daily life.

Before Detroit became the Motor City, it was famous for its wide, tree-lined boulevards and its beautiful parks and was known as the Paris of the Midwest.

When Chauncy Hurlbut died in September 1885, his will provided about $50,000 to be used for beautification of the grounds of Water Works Park and maintenance of his library legacy. The lion's share of his estate, an estimated $250,000—$6 million in 2012— was left to his wife, Philinda Root Hurlbut. When Philinda died within a year of Chauncy, the commissioners put forward a claim to her share of the estate, which was challenged by a group of Hurlbut's relatives.

The will's outstanding probate issues were resolved by 1892, at which time the proceeds originally intended for Philinda were transferred to the board. That year, the first proceeds from the will appeared at Water Works Park in the form of simple floral arrangements. The Hurlbut Memorial Gate made its debut in 1894, followed a year later by the floral clock and a host of increasingly complex floral displays. The public ate it up, and by 1900 Water Works Park was the second-most-popular public park in Detroit after Belle Isle, its riverine sister.

With the exception of closures during World Wars I and II, Water Works Park remained a popular destination. But the Cold War gave national security a new and sinister urgency. When the park closed for the Korean War, it was thought to be permanent. Though a small portion of the park reopened in 1957 in response to popular demand, it was a shadow of its former self and was once more closed to the public by the early 1970s. This time, it was permanent.

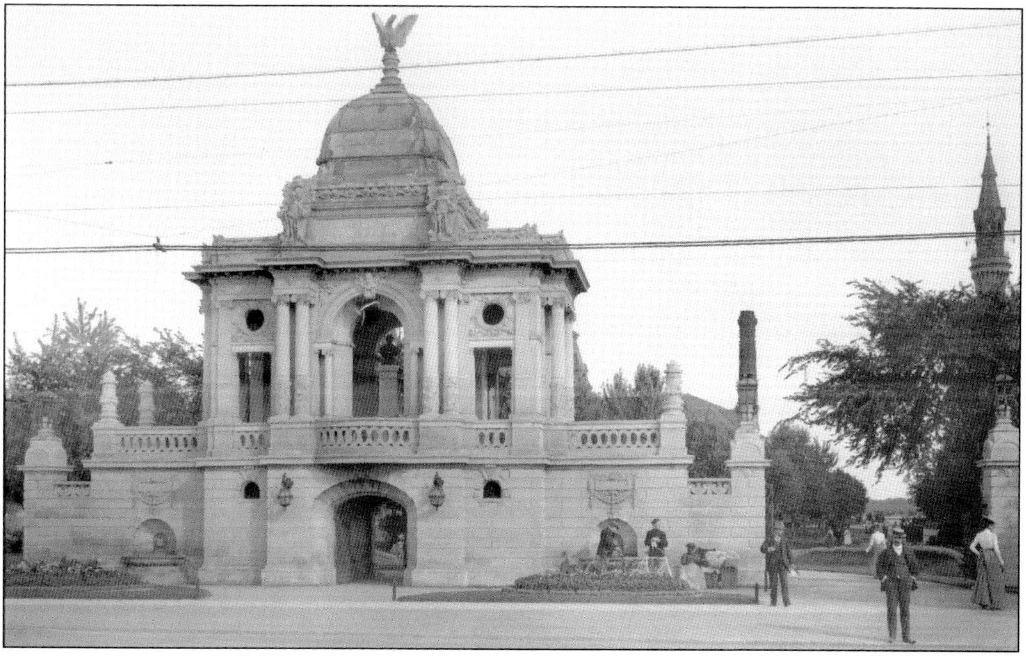

The Hurlbut Memorial Gate, seen in this early view from the 1890s, is a three-tiered triumphal arch. A monument to the park's founder, it was designed by architects Herman A. Brede and Gustave Mueller. It was added to the National Register of Historic Places on March 27, 1975, a year after it was added to Michigan's registry of historic places. (DWSD.)

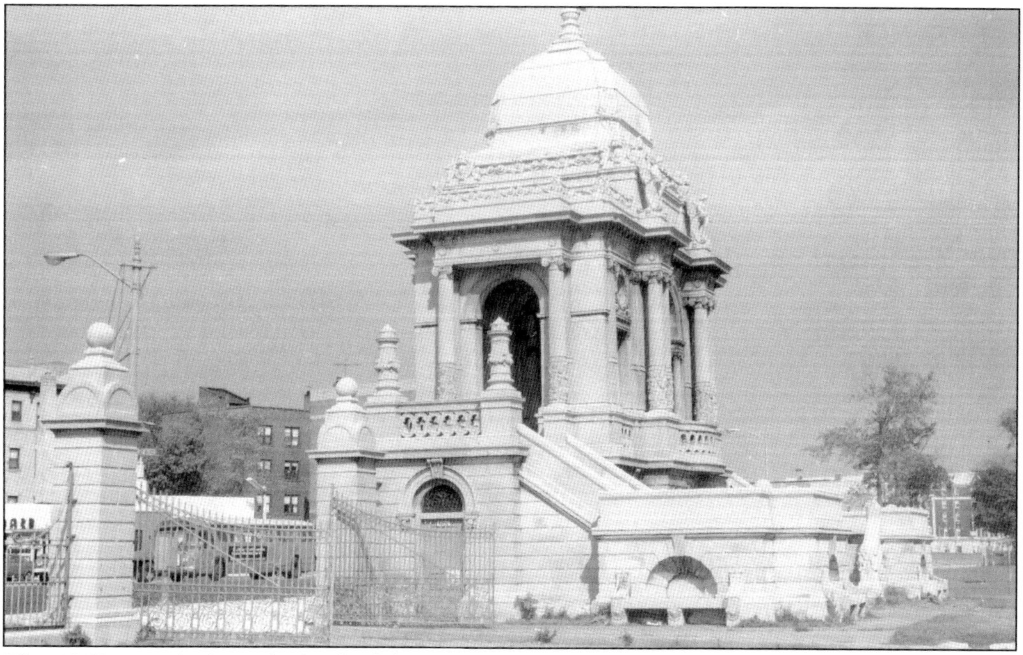

This 1960s view looks at the back of the Hurlbut Memorial Gate from inside the park. In addition to his efforts as president of the Detroit Board of Water Commissioners, the civic-minded Hurlbut was also a member of Detroit's volunteer fire department and the sewer commission. He was also instrumental in the development of Detroit's street railroads. (DWSD.)

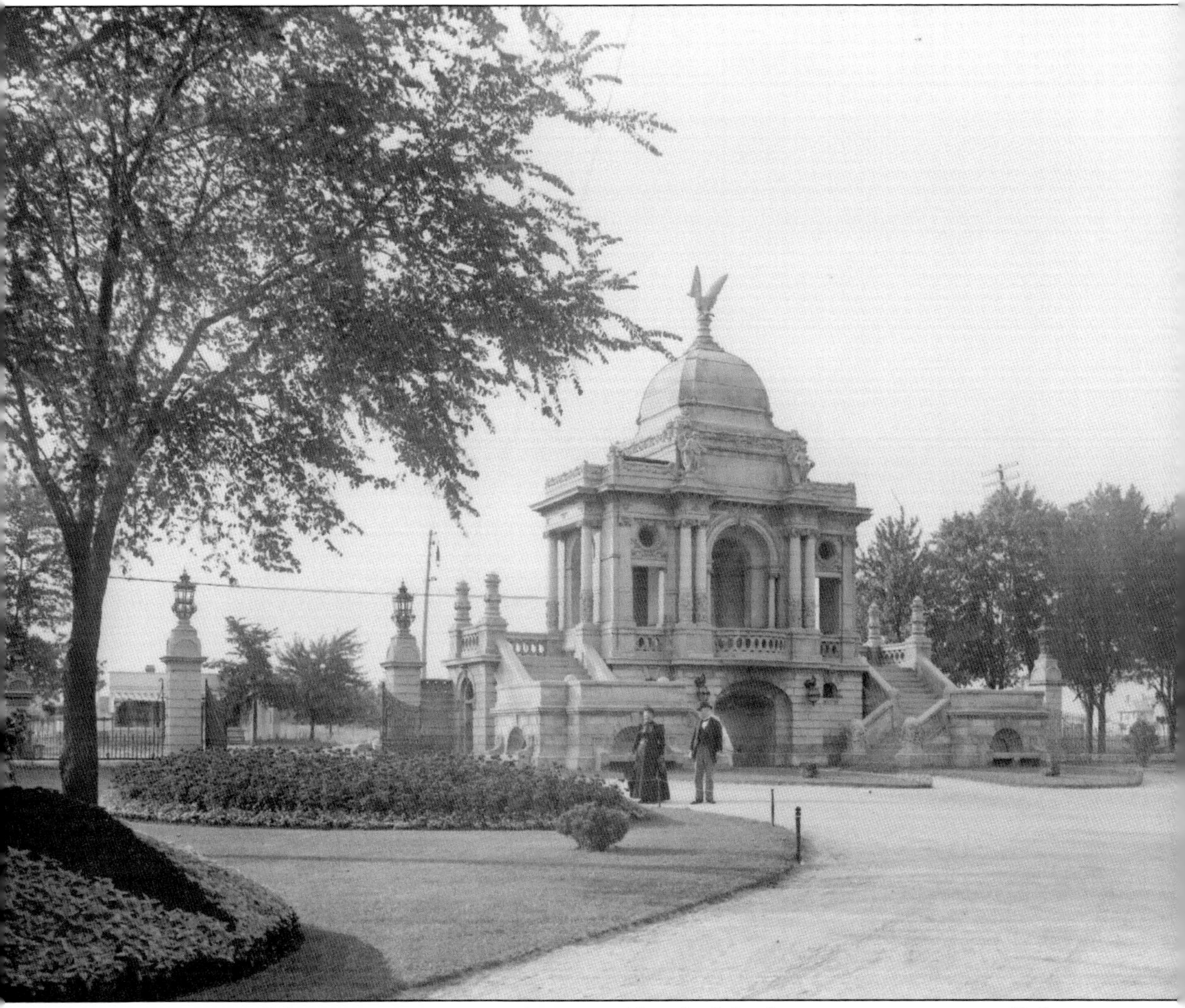

Visitors can ascend the twin staircases to the second level of the limestone monument, 12 feet above the ground. Beneath the dome, a bust of Hurlbut used to sit on an Aberdeen granite pedestal. Unfortunately, the brass bust was stolen by vandals in 1975. (DPC/LOC.)

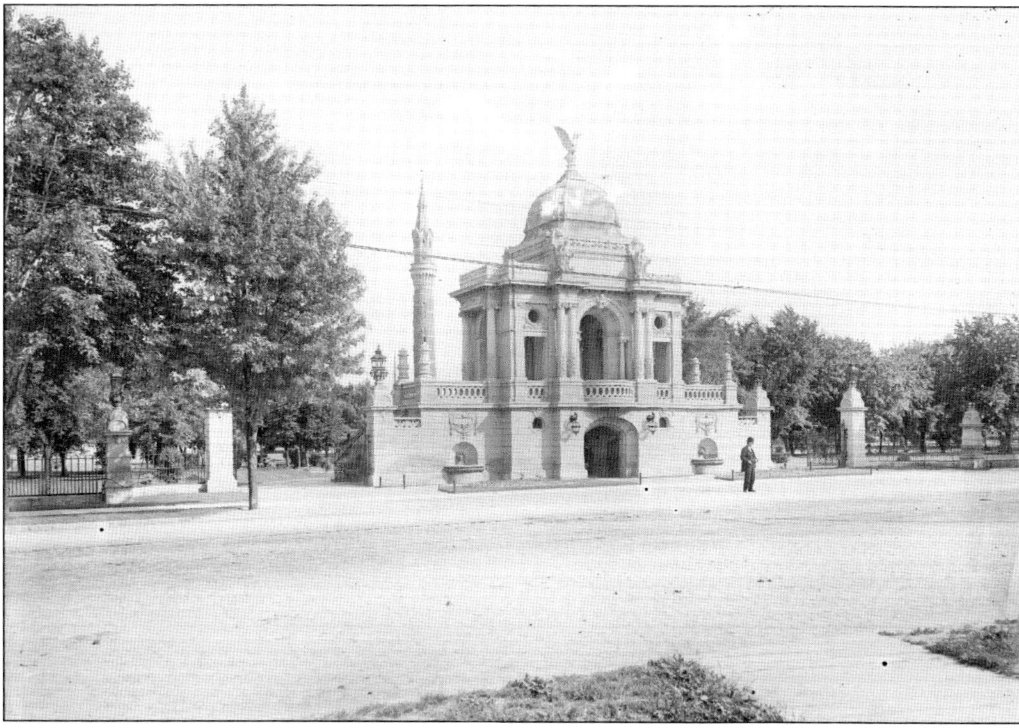

The grand Hurlbut Memorial Gate is shown above from the Jefferson Avenue side in the 1890s. The rear view of the gate seen below, looking north from inside the park, is thought to be from the same era. The gate is a Beaux Arts structure, an architectural style popular in Victorian times. A victory eagle is proudly perched on top of the dome, which rests on columns. The monument is 132 feet wide, 40 feet deep, and 50 feet high. It cost $30,000 to build, an amount equivalent to nearly $770,000 in 2012. (Above, DPC/LPC; below, DWSD.)

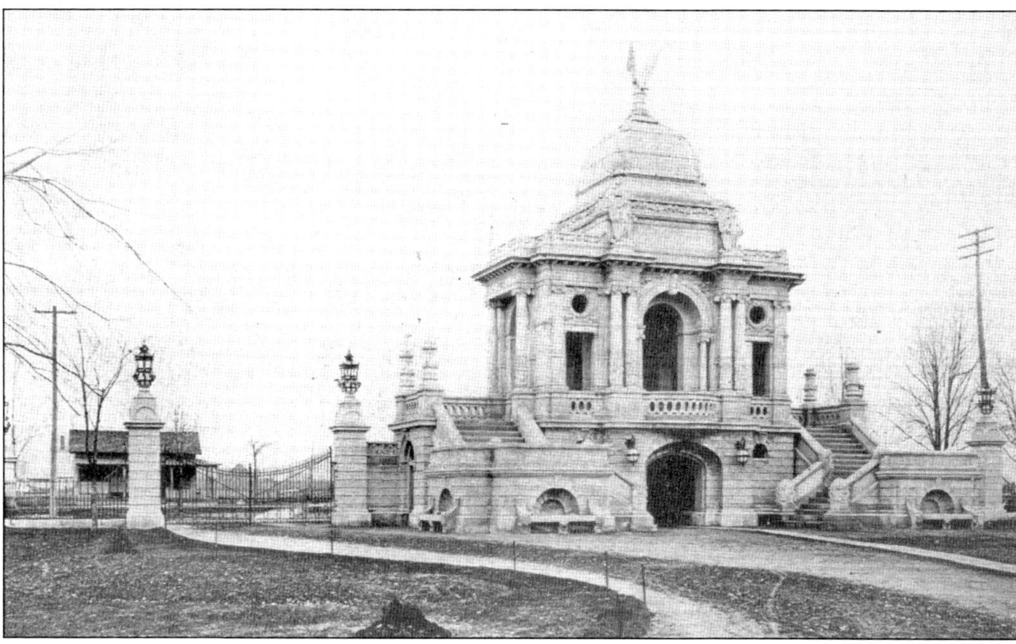

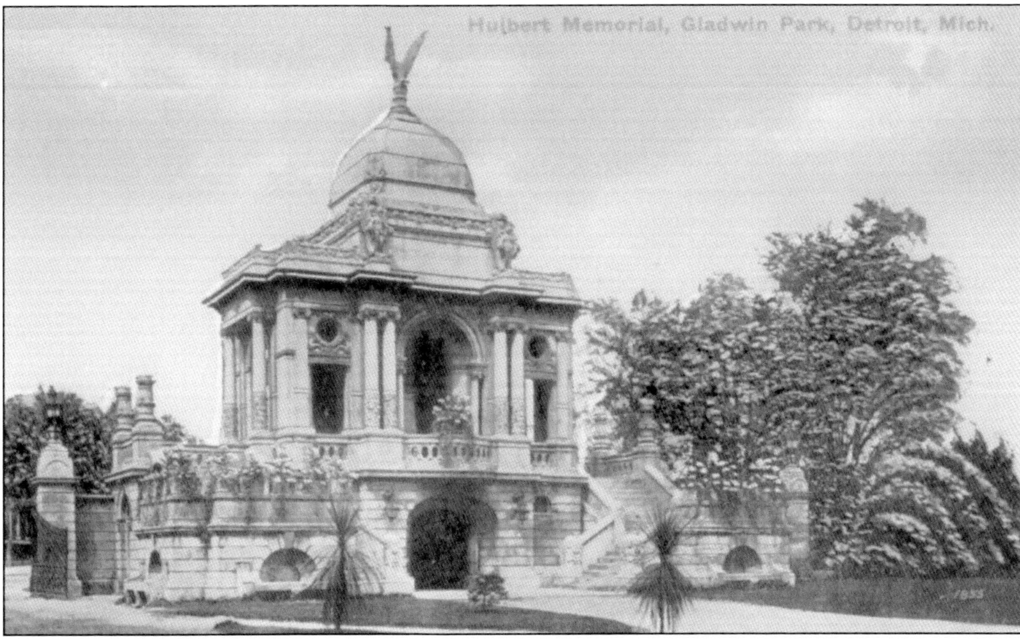

Here, the gateway is decked out in its holiday finest, perhaps for Easter. This image probably dates from around 1900. The view of the gate is seen from inside Water Works Park, looking north to East Jefferson Avenue. (Author.)

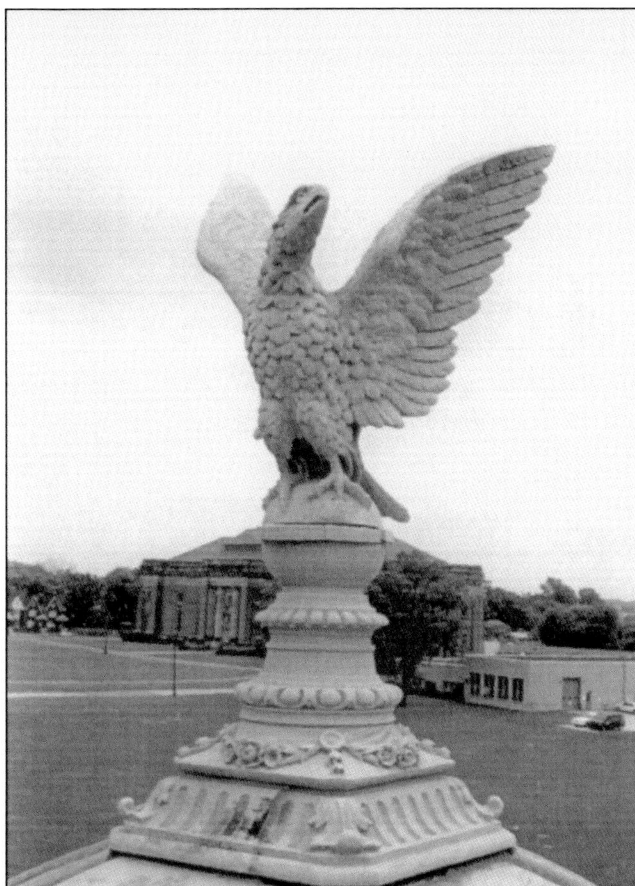

The majestic victory eagle on top of the dome is shown here in 1997, prior to the start of renovations on the century-old monument. Contracts worth nearly $1 million were approved to rehabilitate the aging Hurlbut Memorial Gate and the old Ladder Company No. 14 fire station on East Jefferson Avenue. (DWSD.)

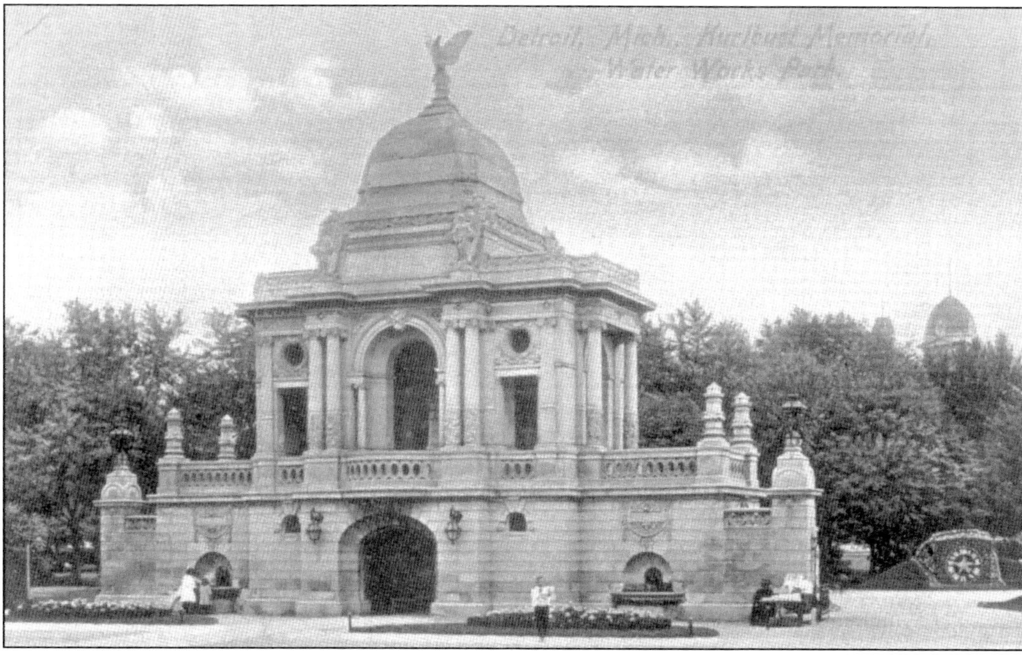

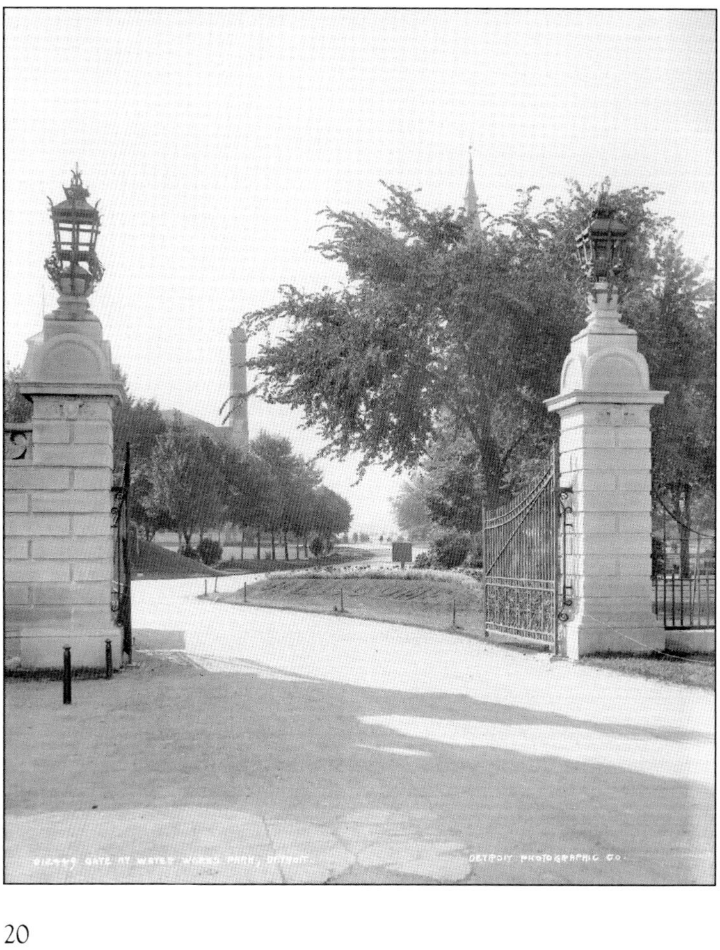

The Hurlbut Memorial Gate and floral clock were featured on many popular postcards of Water Works Park in the first decade of the 20th century. (Author.)

A view of the landscaped grounds, pumping station, and water tower is seen through the gateway just west of the Hurlbut Memorial Gate in the 1890s. To the left, partially obscured by the iron gate, is the floral clock. (DPC/LOC.)

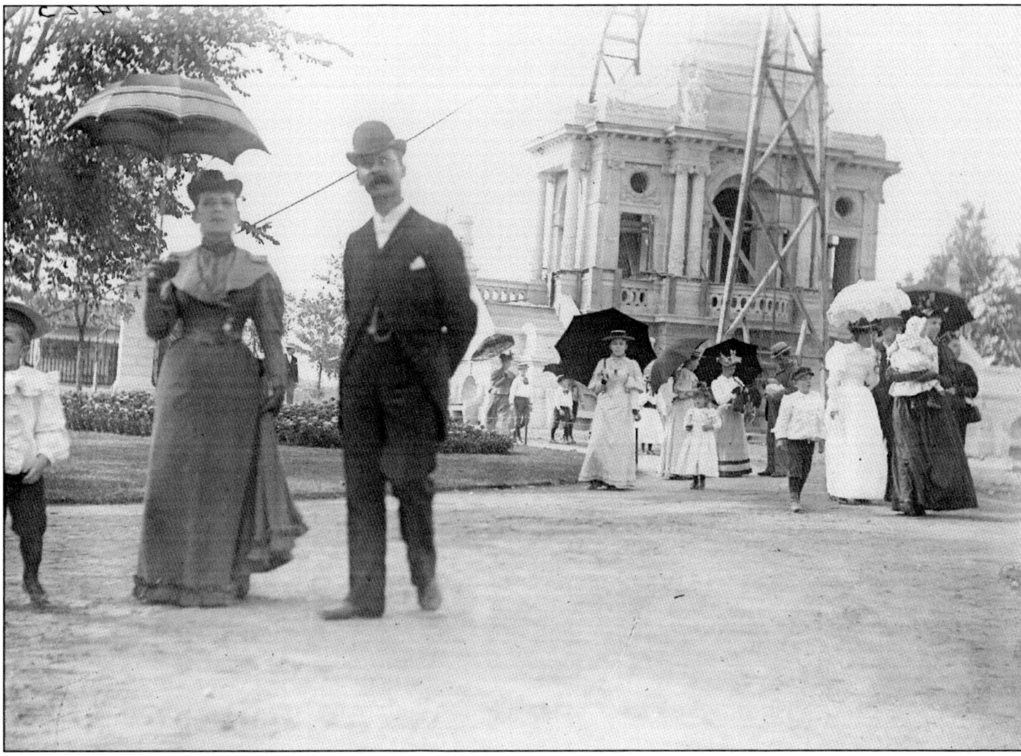

Visitors pass the Hurlbut Memorial Gate as they enter Water Works Park. The memorial, then under construction, was designed by architects Herman Bede and Gustave Mueller and hewn from Bedford limestone. It honors Chauncy Hurlbut, a leader of Detroit's Board of Water Commissioners and father of Water Works Park. Adjusted for inflation, the $30,000 cost to build in 1893 is equivalent to more than $740,000 in 2011. (Burton Collection.)

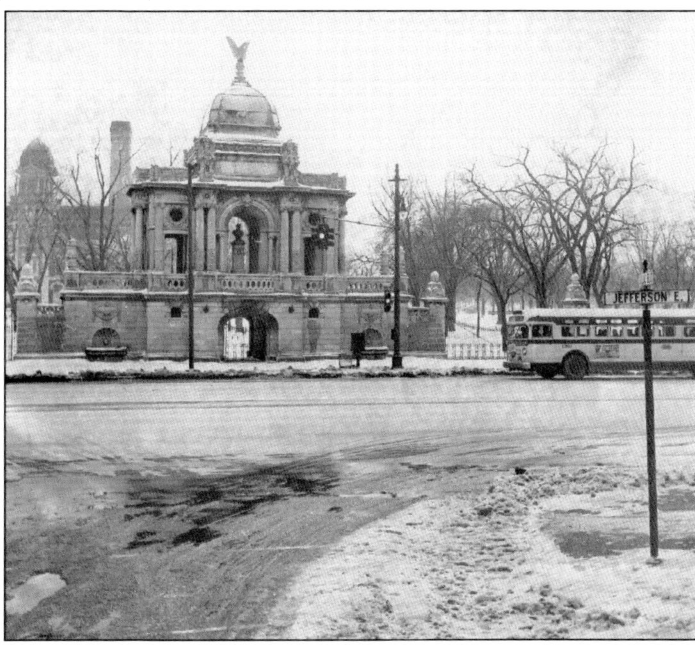

A bus has stopped at the light in front of the Hurlbut Memorial Gate on a wintry day in the 1950s. From its earliest days, Detroiters had ready and easy access to Water Works Park by means of both streetcars and buses. (Author.)

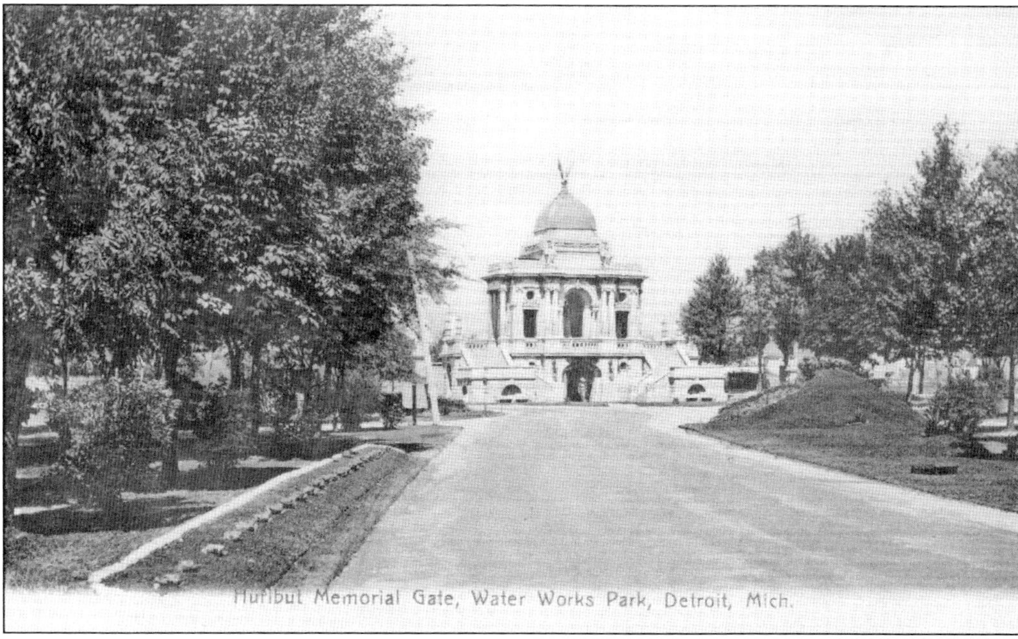

Most images of Water Works Park from its many years as a public gathering space tend to show the grounds and exquisite floral displays as finished works. In most, workers and their tools are nowhere to be seen, but in this unusual postcard image of the back of the Hurlbut Memorial Gate from around 1908, a ladder is propped on a tree to the left. (Author.)

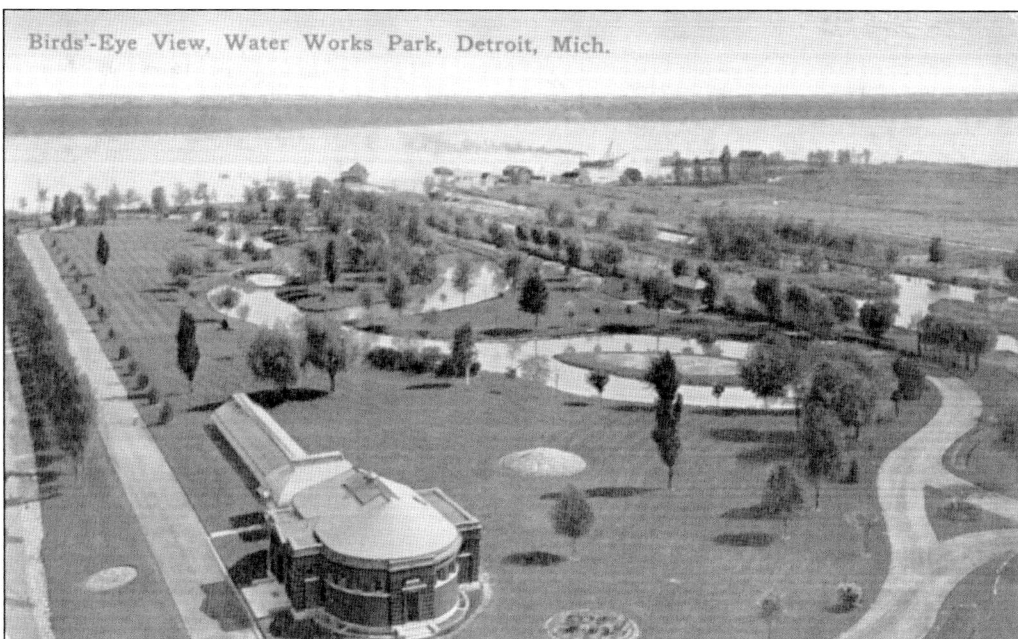

This bird's-eye view of Water Works Park from around 1905 looks out from the observation platform near the top of the water tower. The Hurlbut Library, created to house the personal collection of Chauncy Hurlbut, is in the foreground. Beyond it is the winding canal. In the distance are the Detroit River and Belle Isle. (Author.)

After spending a pleasant day on the grounds of Water Works Park in the 1890s, a group of Detroiters, picnic baskets in hand, approach the Hurlbut Memorial Gate and East Jefferson Avenue to catch a ride home on one of Detroit's efficient streetcars. A close look at this photograph shows that it has been reversed. (DPC/LOC.)

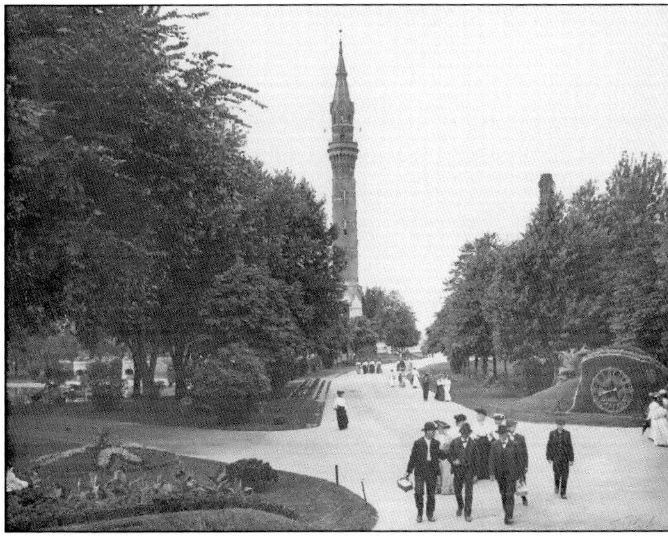

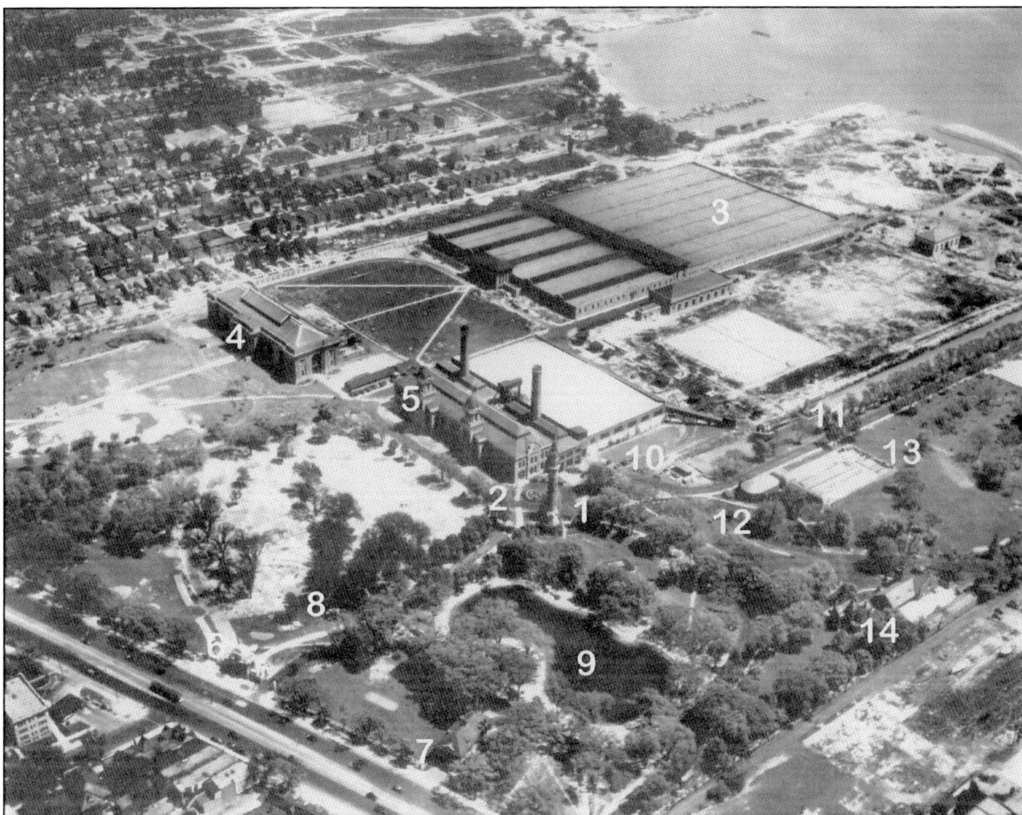

This aerial view of the park from 1930 clearly specifies all the nearby landmarks. Visible are the standpipe (1), the floral calendar (2), Detroit's first water treatment plant (3), and both pumping stations (4 and 5). By this time, the pumping stations were functioning as treated-water discharge, or "high-lift" pump stations. Also visible are the Hurlbut Memorial Gate (6), Ladder Company 14 (7), the floral clock (8), lake (9), pavilion (10), coaling canal (11), Hurlbut Library (12) with greenhouses (13), and the civil engineer's house and stables (14). (*Detroit News*.)

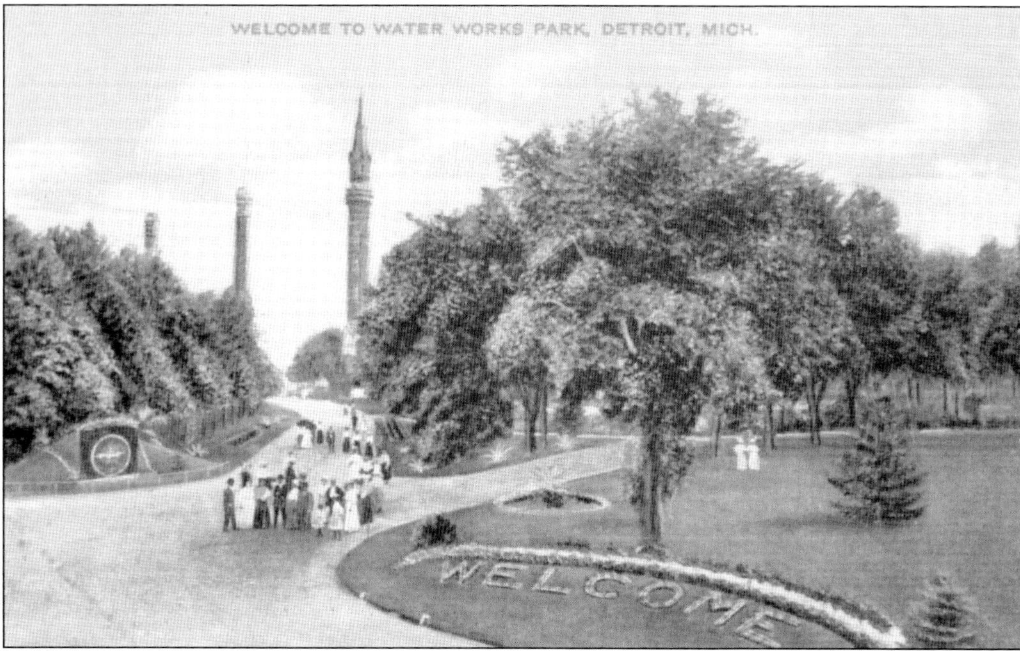

Upon entering Water Works Park in the early years of the 20th century, visitors found themselves surrounded by a lushly manicured green space full of stately structures. The hand-tinted image from a postcard (above) is what a visitor would have seen around 1900. The postcard image below is from after 1910, when the members of the Detroit Common Council renamed the site Gladwin Park. The park's namesake, Maj. Henry Gladwin, was the British commanding officer of Fort Detroit in 1763 during the months-long siege by a coalition of as many as 900 Native American warriors led by Pontiac, an Ottawa chief also known as Obwandiyag. (Both, author.)

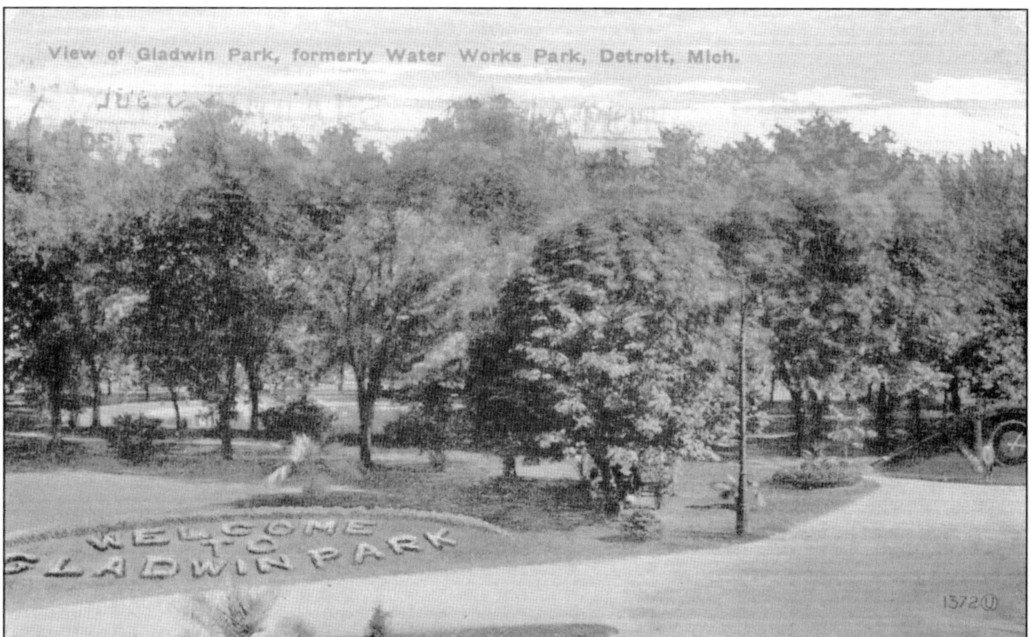

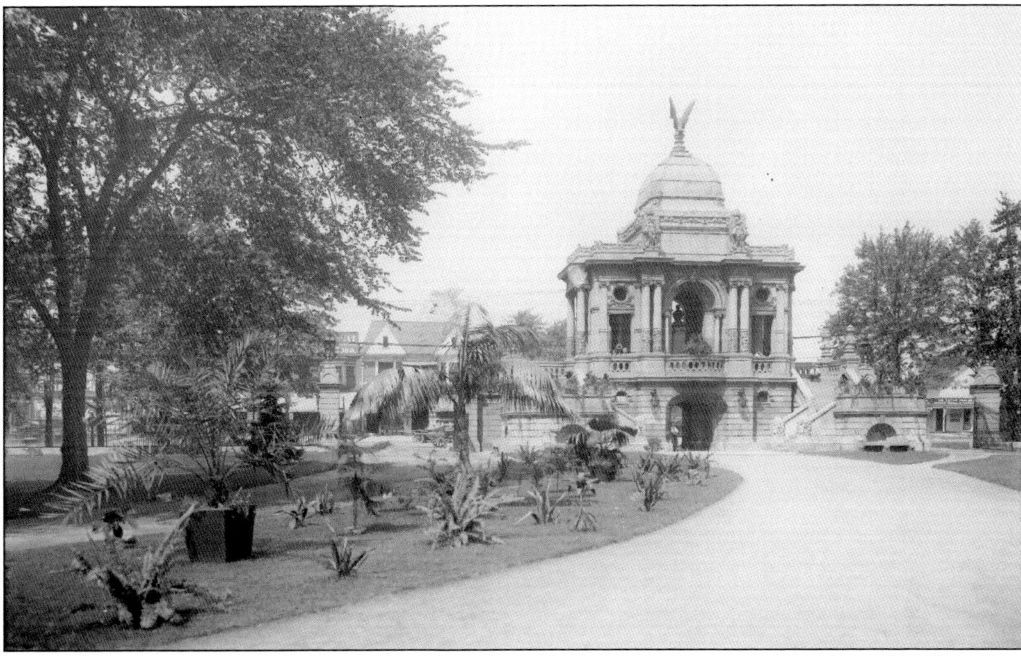

This image of Water Works Park from the 1890s shows an early variation of the floral displays for which the park became so renowned. The first displays appeared in 1892 following the end of probate of the Hurlbut will. The area around the ice cream soda shop across the street is now occupied by Pewabic Pottery. (DPC-LOC.)

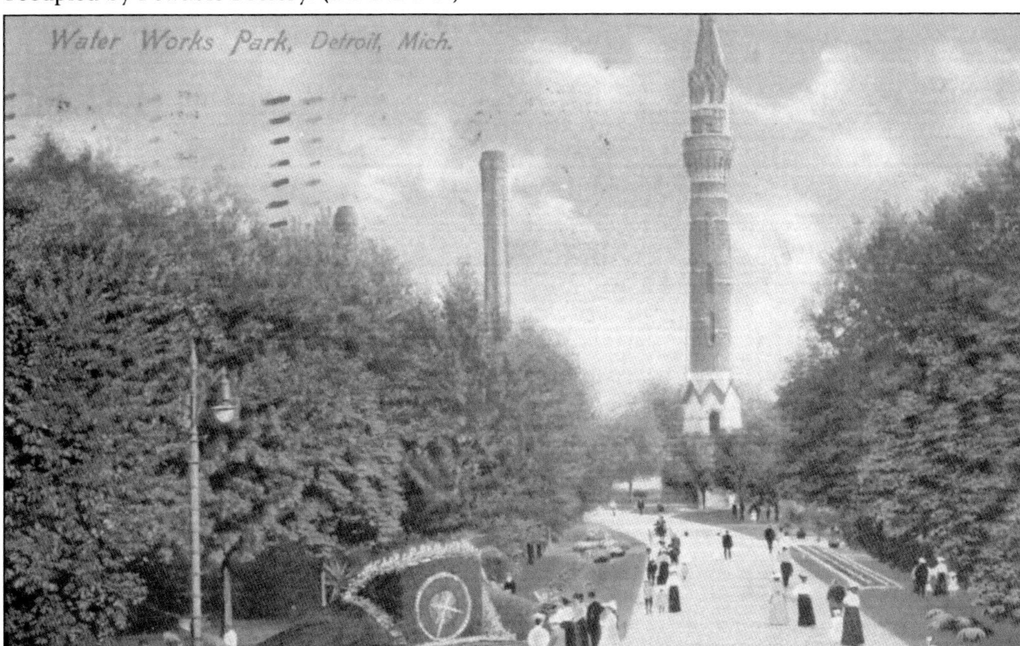

This classic, hand-tinted postcard of the park appears to be from about 1910 and looks toward the Detroit River from the Hurlbut Memorial Gate. Visitors in their Sunday best leisurely admire the world-famous floral clock and intricate floral creations bordering the main pathway into the park. (Author.)

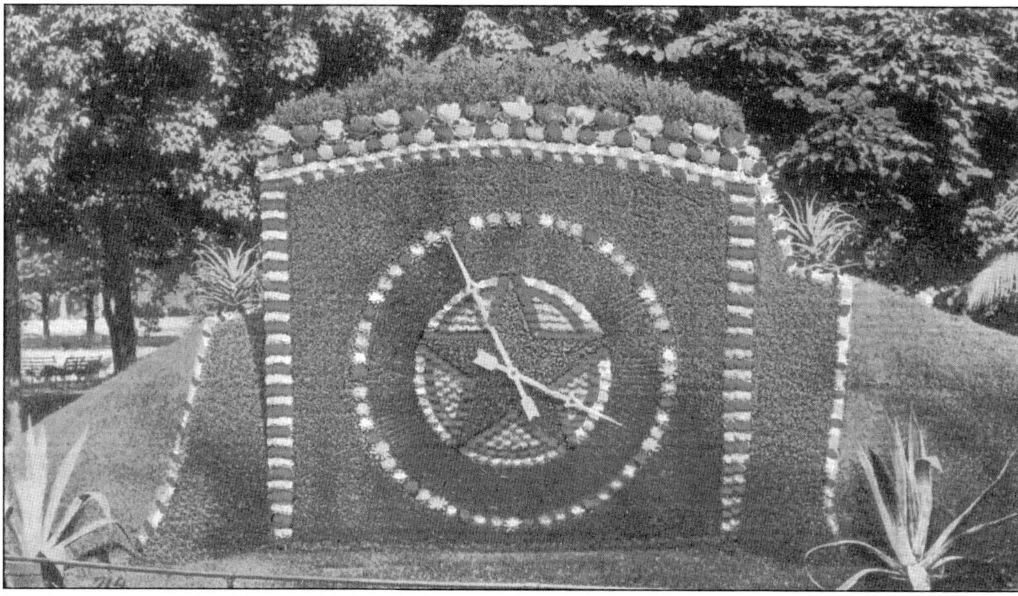

The floral clock was one of the signature attractions at Water Works Park for many years, and a popular subject for postcards. The clock stood 7.5 feet high and 10 feet wide, with a face 6 feet in diameter. It was hydraulically powered by cup-shaped paddle wheels, which extended into the water main beneath the knoll on which it sat. The clock was removed in 1934 because of poor timekeeping and high maintenance costs. It was then acquired by Henry Ford, who displayed it at Greenfield Village from 1935 to 1974. It was returned to the Detroit Water and Sewerage Department in 1989 and is currently in storage. (Both, author.)

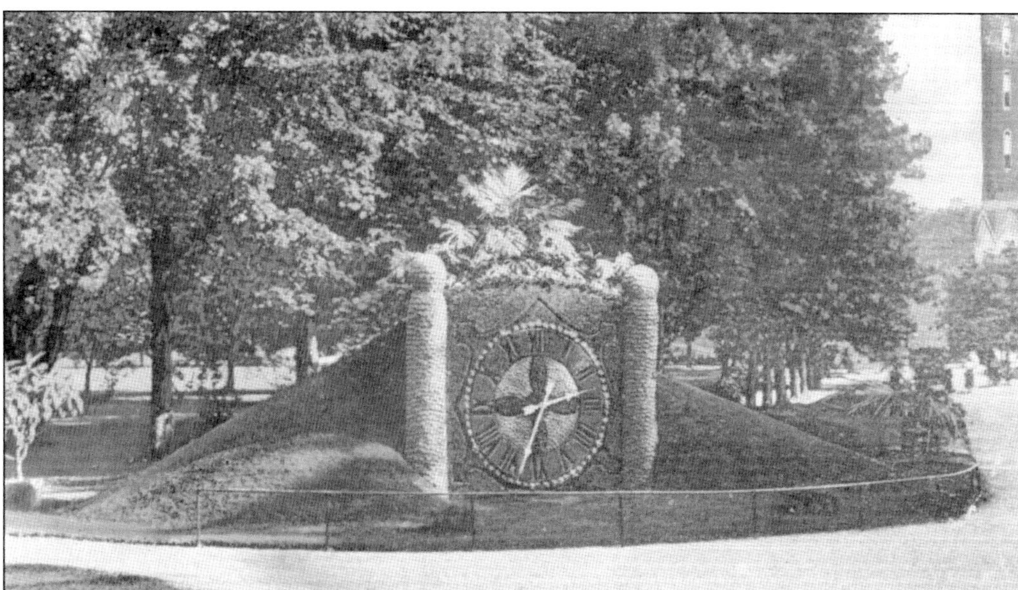

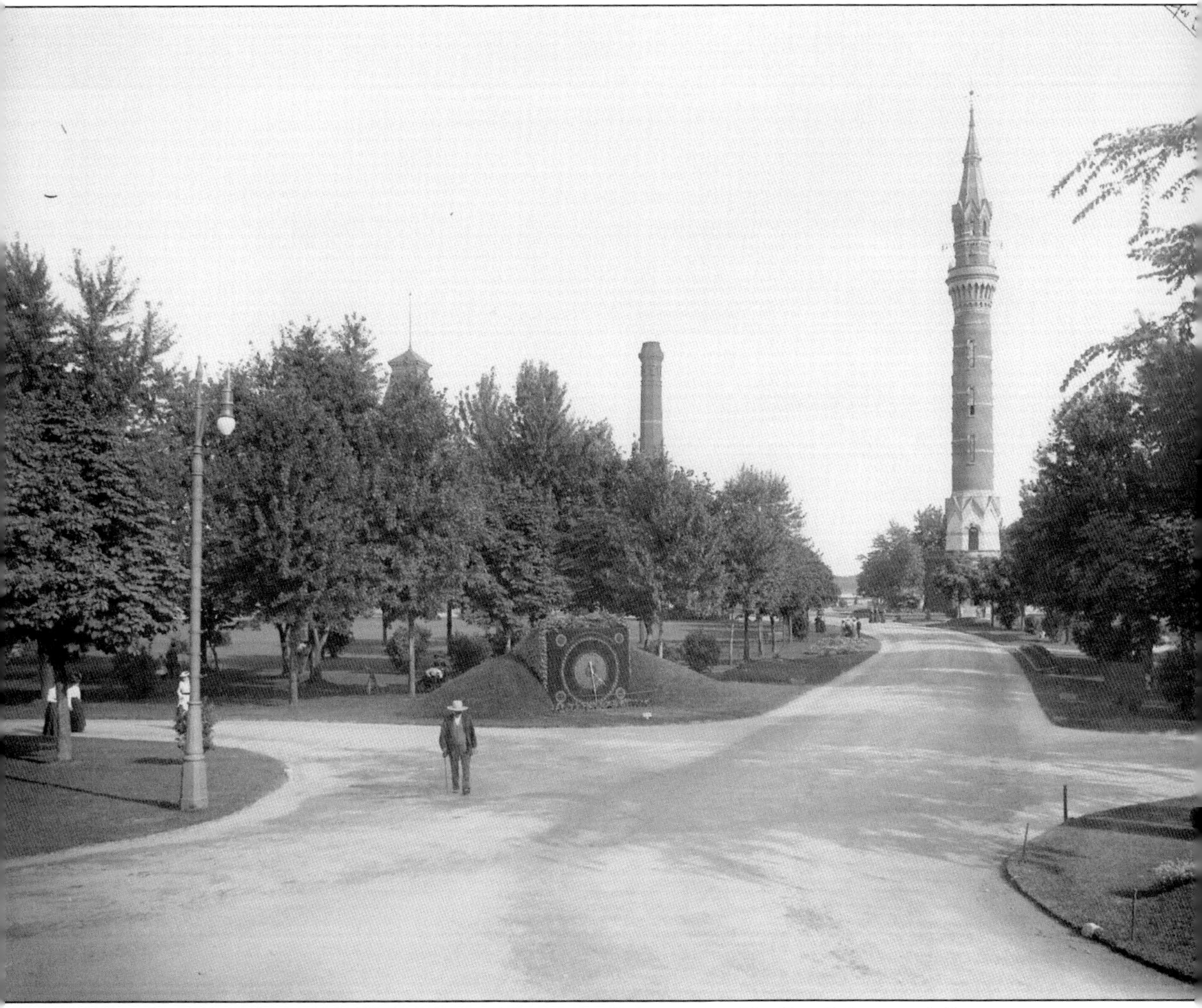

Detroiters enjoy the peaceful serenity of the grounds near the Hurlbut Memorial Gate entrance in the 1890s. The famous floral clock is at the crossroads in the foreground, and the water tower can be seen in the distance. The roof of the original pumping station is on the left, just above the trees. (DPC-LOC.)

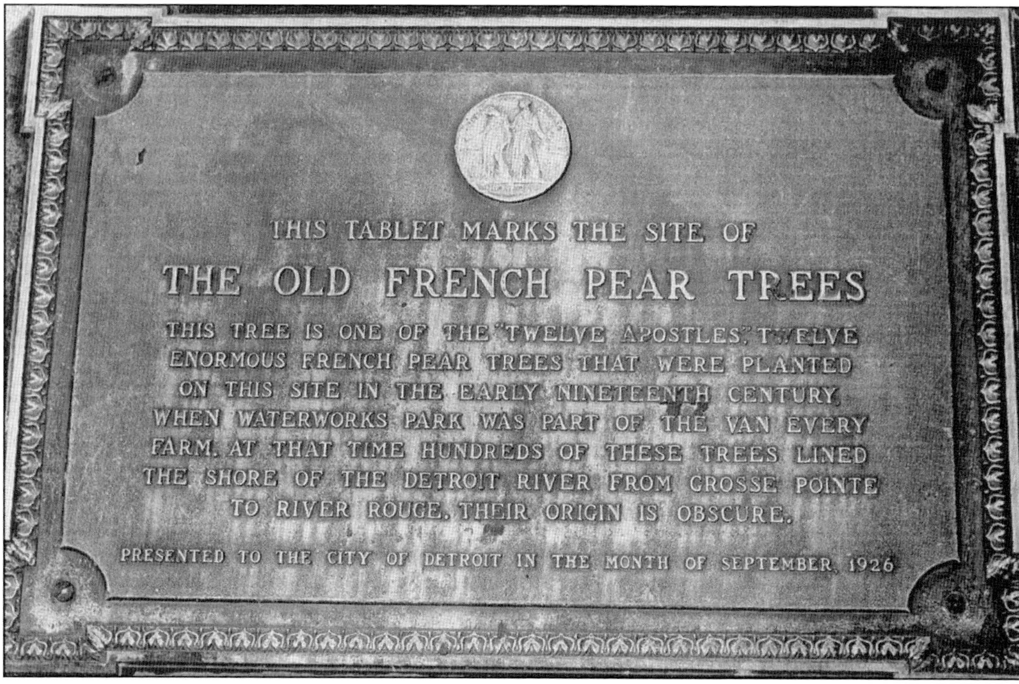

The brass plaque that once sat among the Twelve Apostles in Water Works Park is shown in a close-up and mounted on a boulder on the grounds. Most of the information on the plaque is accurate, but the trees were actually planted by settlers during the French occupation of Detroit in the early 18th century. Once abundant along the Detroit River, the trees produced enormous amounts of fruit. The fruit was prized not for its food but as an essential ingredient in perry, a type of pear cider still popular in parts of Britain and in the Normandy region of France. The current whereabouts of the plaque are unknown. (Both, DWSD.)

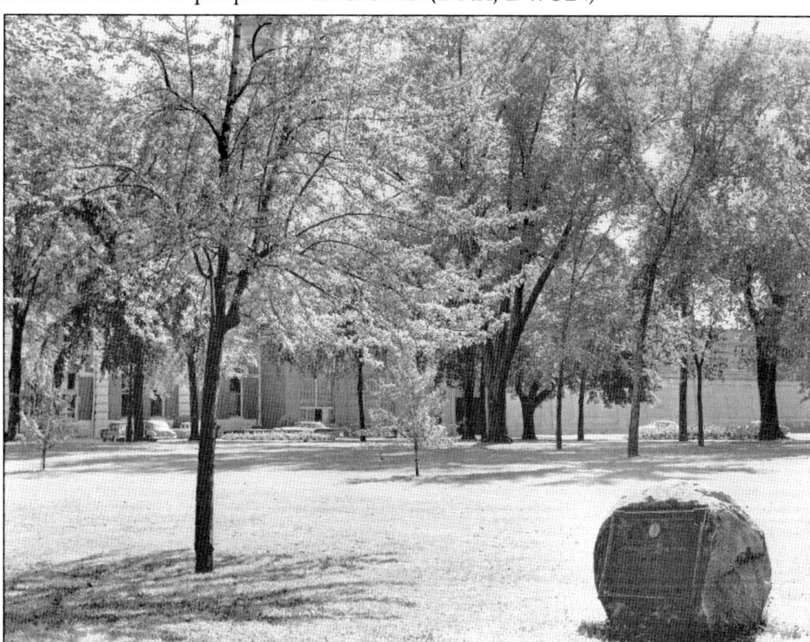

These two views from 1938 show Judas, the last surviving member of the Twelve Apostles. The French pear trees, also known as Mission or Jesuit pears, were planted in a circle, according to local custom, in the vicinity of the original pumping station, near the site of the current administration building. In accordance with Biblical accounts, Judas was planted apart from the other 11. Badly scarred from countless lightning strikes and liberally patched with cement in an effort to extend its life, the 200-year-old, 63-foot-tall giant succumbed to old age in the 1940s. (Both, author.)

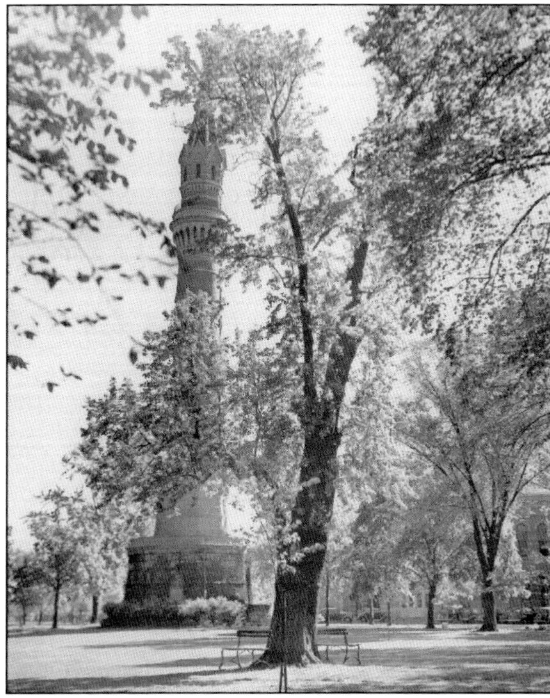

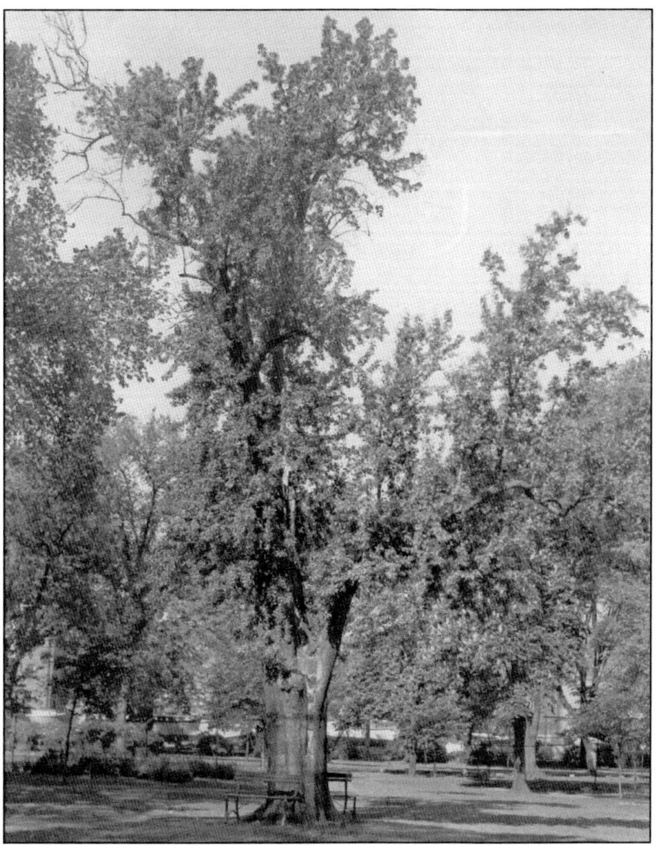

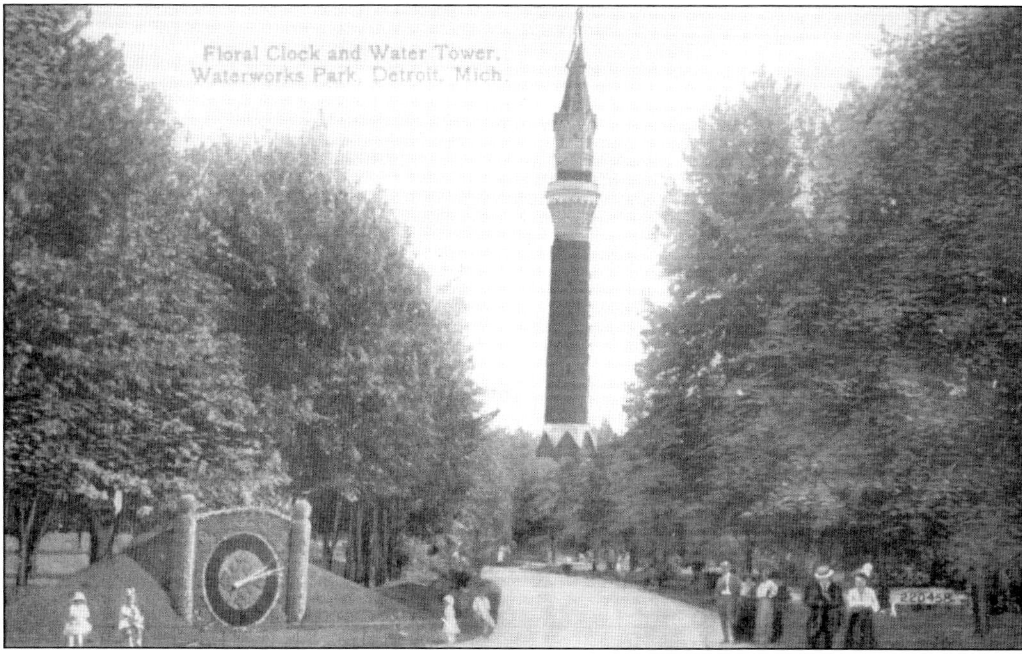

This hand-tinted color postcard of the park's entryway shows the popularity of the park at the time. The more thickly forested park dates this photograph to around 1900. (Author.)

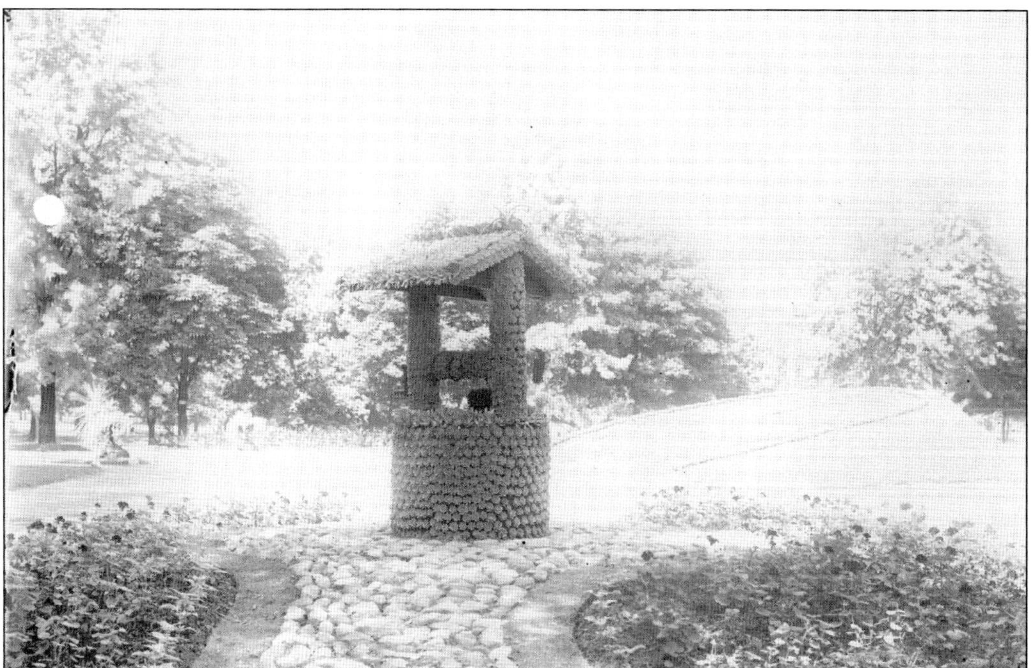

Water Works Park was known for the variety and ingenuity of its floral displays, including such things as a cow in a field of corn, and a teepee with smoke venting from the top. This wishing well near the floral calendar and standpipe was created by groundskeepers. (DWSD.)

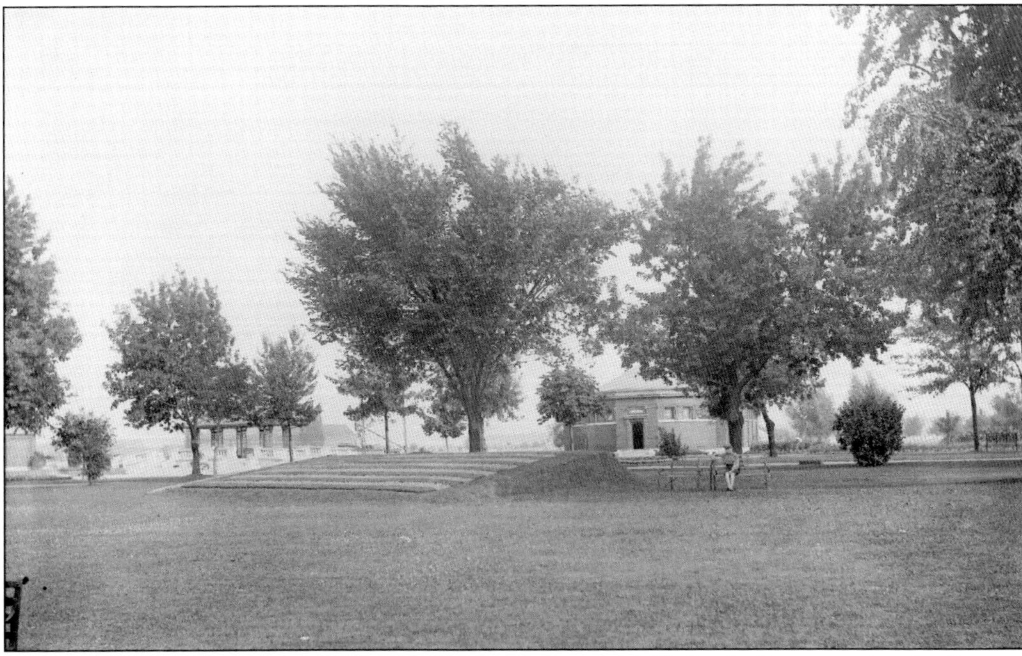

Detroit the City of the Strait, published in 1897 by Rand McNally & Company for the General Passenger Department of the Michigan Central Railroad, gives this description of the meticulously sculpted grounds of Water Works Park: "The extensive grounds with their velvety lawns, shady walks, placid lagoons, splashing fountains, bright flower beds and memorial gateway, make it by no means the least important of Detroit's many beautiful parks." The author of that passage was no doubt inspired by views such as the floral flag on a gentle rise in the vicinity of the pavilion and the Hurlbut Library (above), and the floral splendor of the grounds near the Hurlbut Memorial Gate, looking west to the lake and fountain, below. (Both, DPC/LOC.)

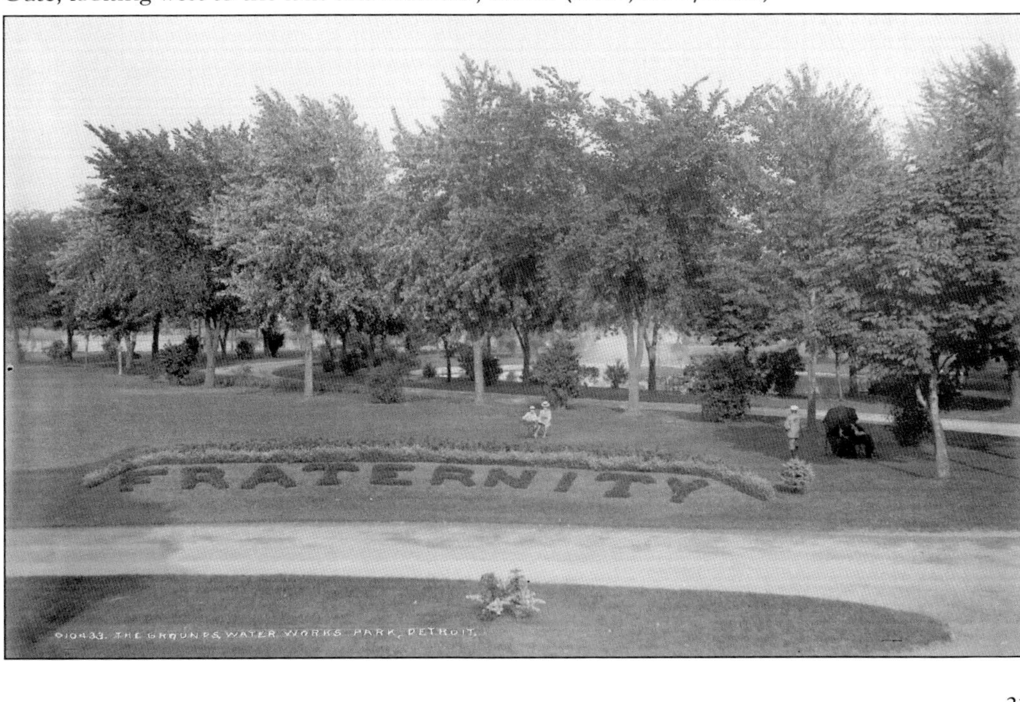

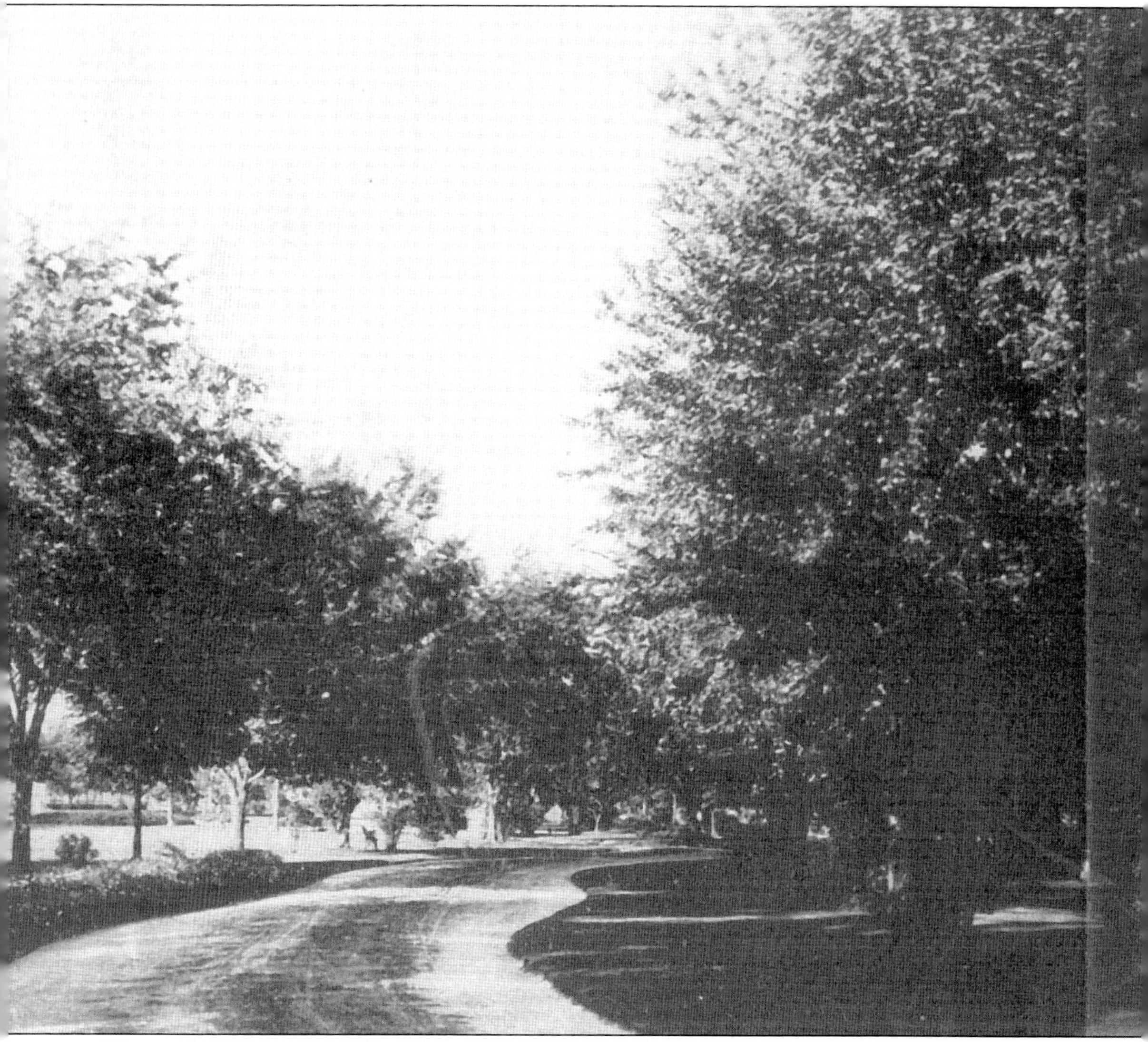

This mid-park panorama from the 1890s gives a good perspective of the way Water Works Park would have been experienced by visitors. At the extreme left is the eastern end of the Hurlbut Memorial Gate. Beyond the trees to its right is Jefferson Avenue. Although it is difficult to

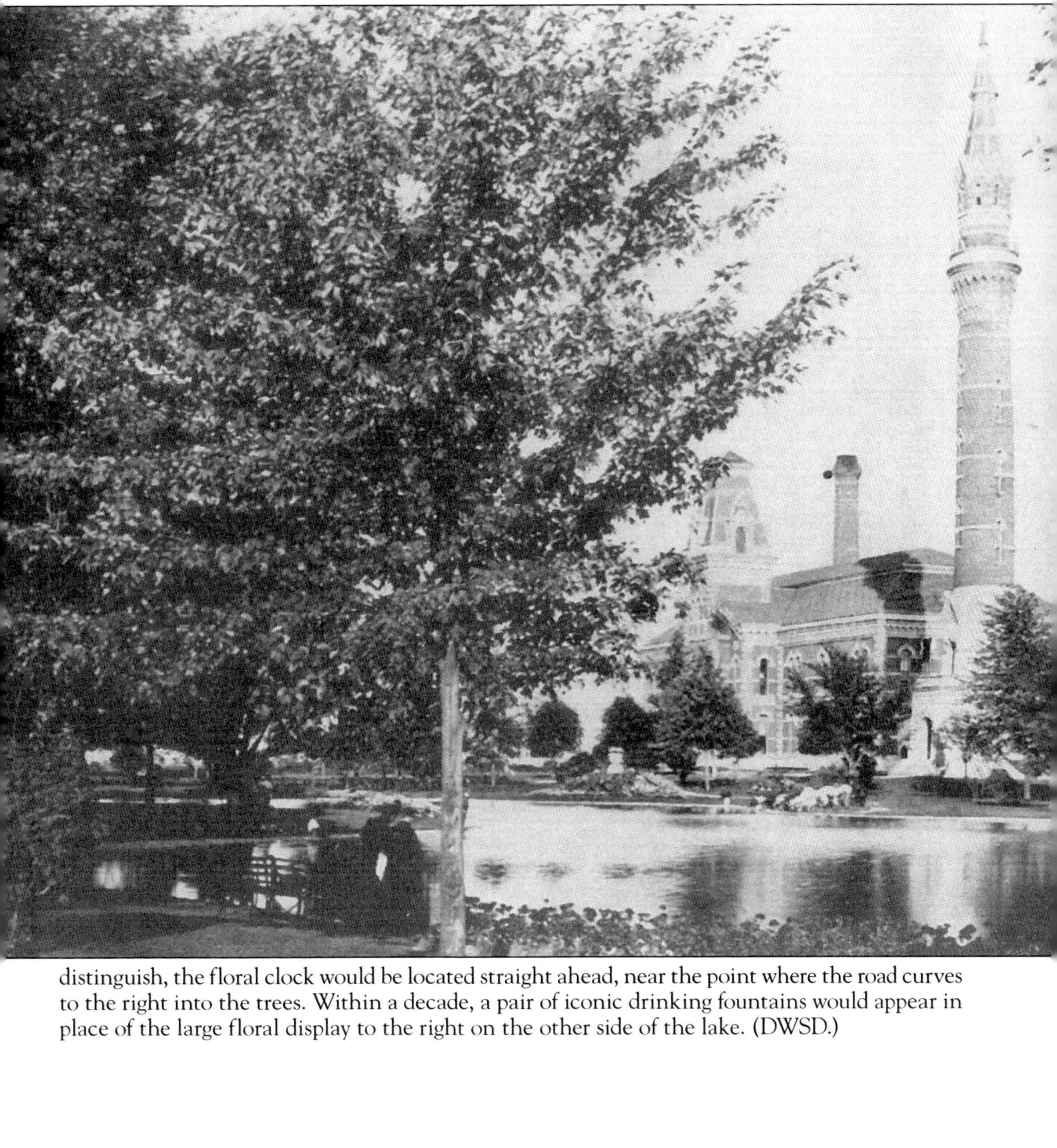

distinguish, the floral clock would be located straight ahead, near the point where the road curves to the right into the trees. Within a decade, a pair of iconic drinking fountains would appear in place of the large floral display to the right on the other side of the lake. (DWSD.)

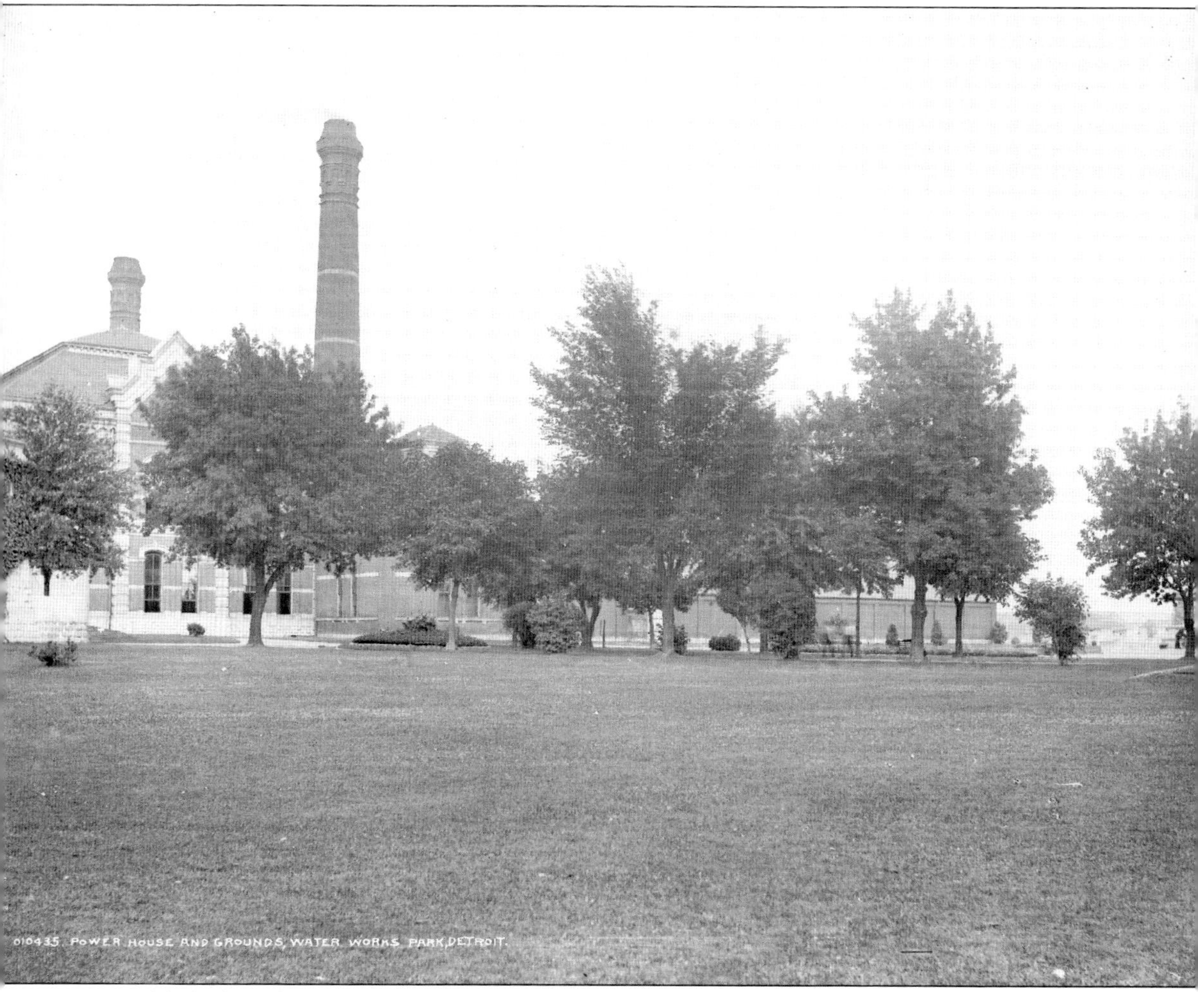

It is likely that this group of trees, in the vicinity of the original pumping station in the 1890s, includes members of the Twelve Apostles. (DPC/LOC.)

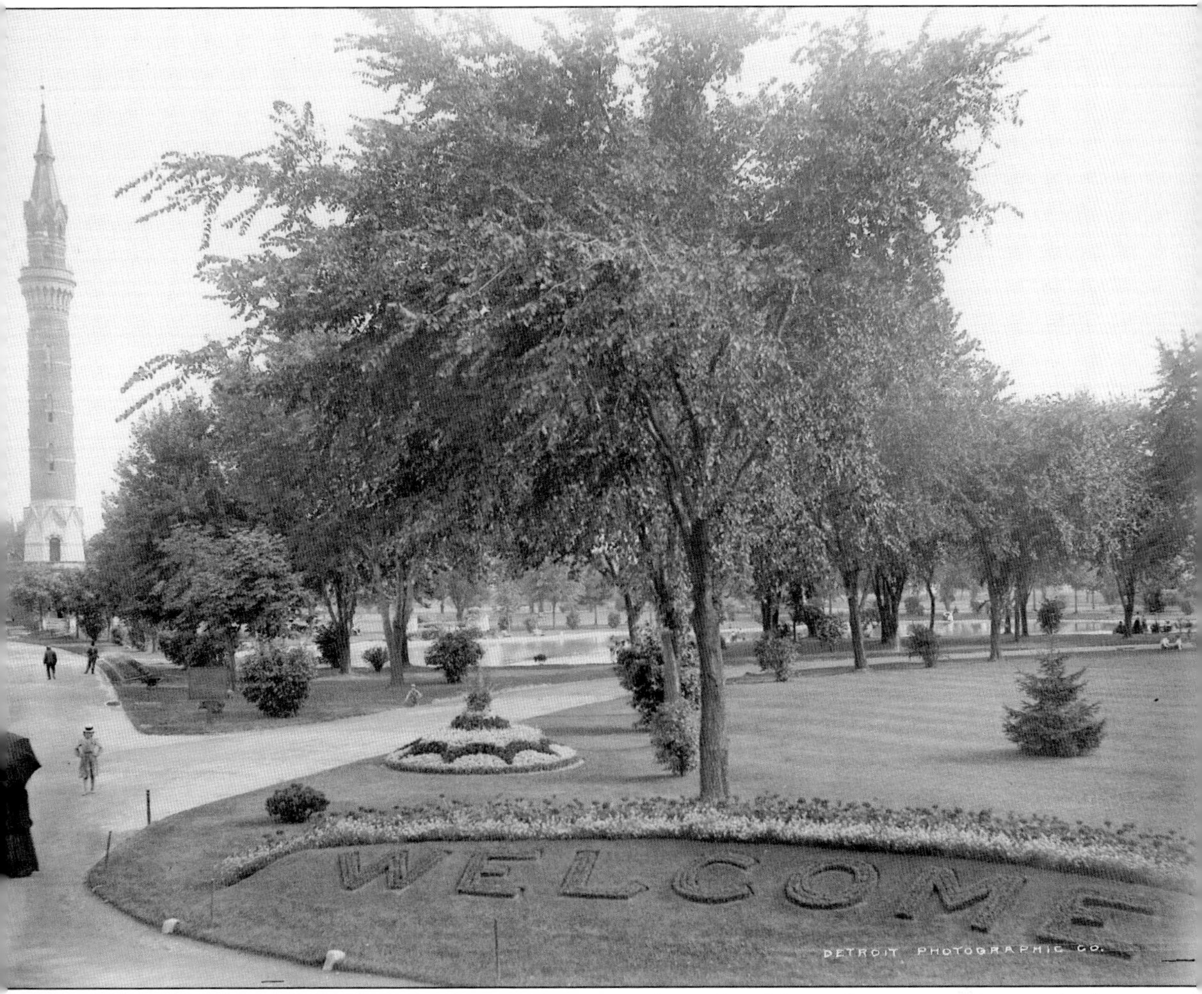

This 1890s photograph, taken from the front entrance, shows the elaborately manicured grounds of Water Works Park from the front entrance. Members of the grounds crew utilized two greenhouses to maintain existing displays and to create new ones. (DPC/LOC.)

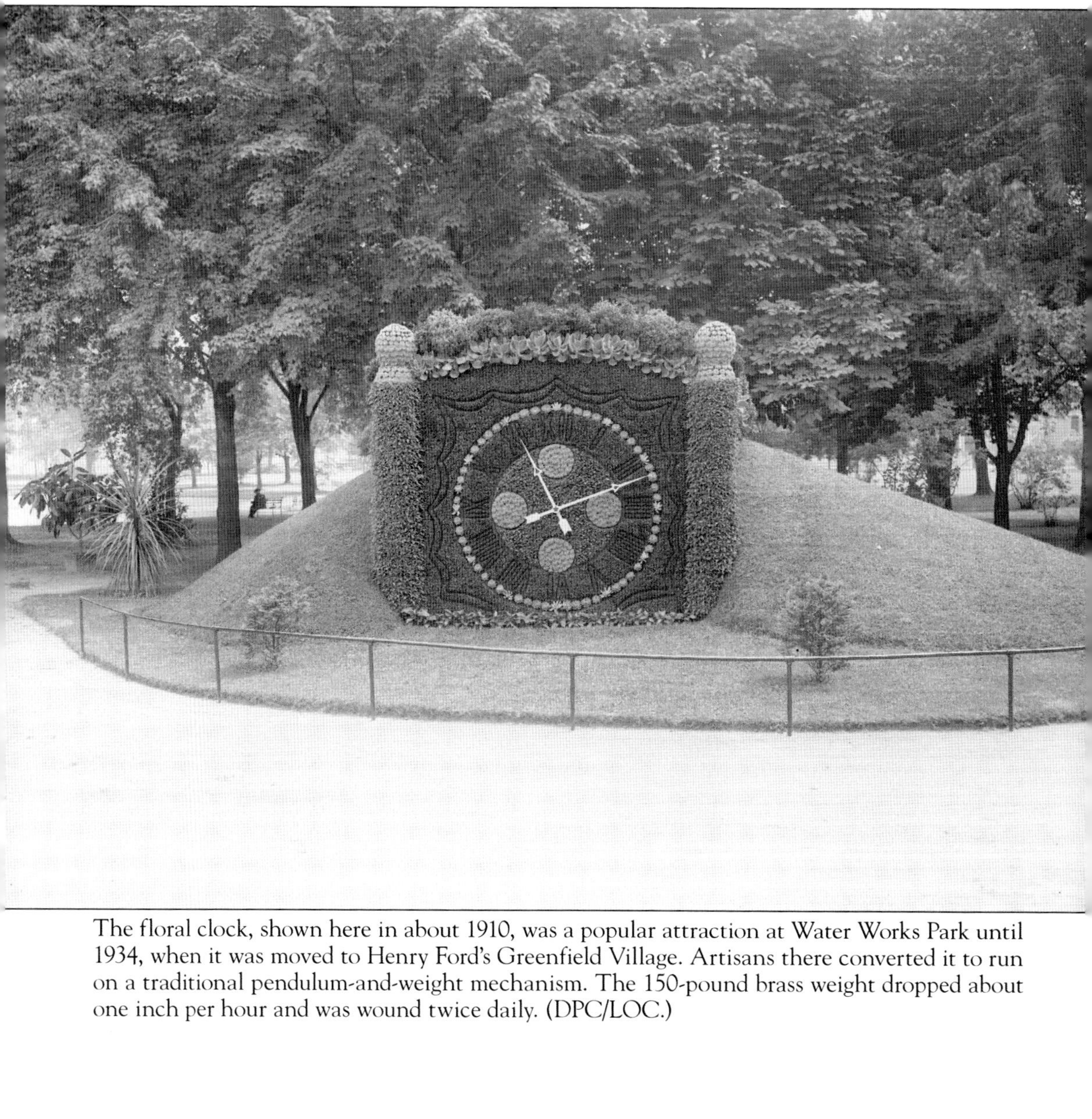

The floral clock, shown here in about 1910, was a popular attraction at Water Works Park until 1934, when it was moved to Henry Ford's Greenfield Village. Artisans there converted it to run on a traditional pendulum-and-weight mechanism. The 150-pound brass weight dropped about one inch per hour and was wound twice daily. (DPC/LOC.)

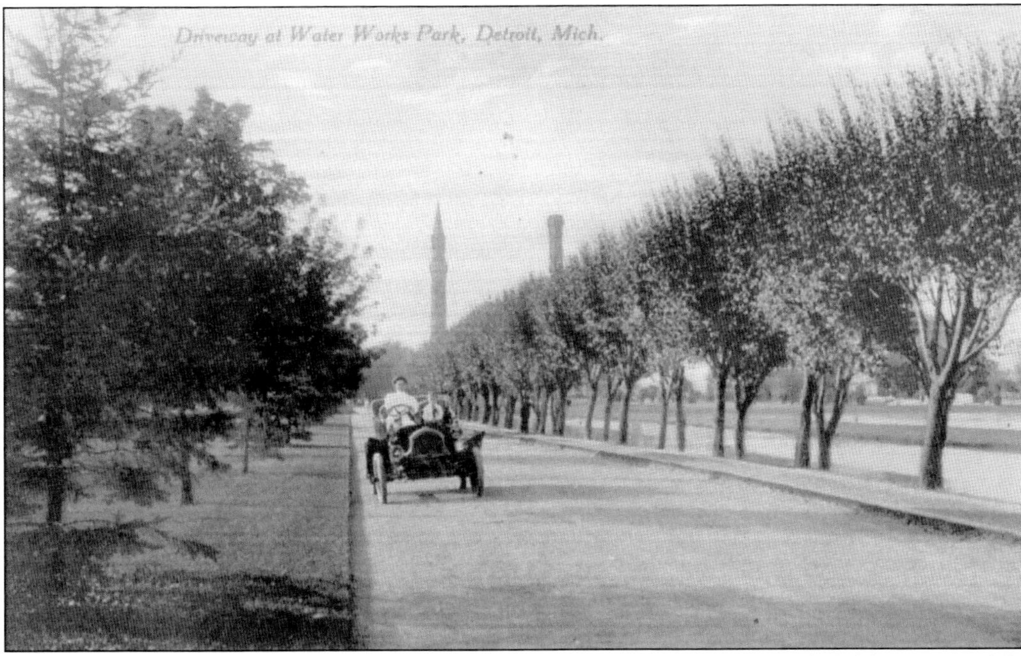

This image from the first decade of the 20th century shows a mother and child, in one of the newfangled horseless carriages made in Detroit, enjoying a drive—perhaps on a Sunday in summer—down the tree-lined driveway along the coaling canal in Water Works Park. (Author.)

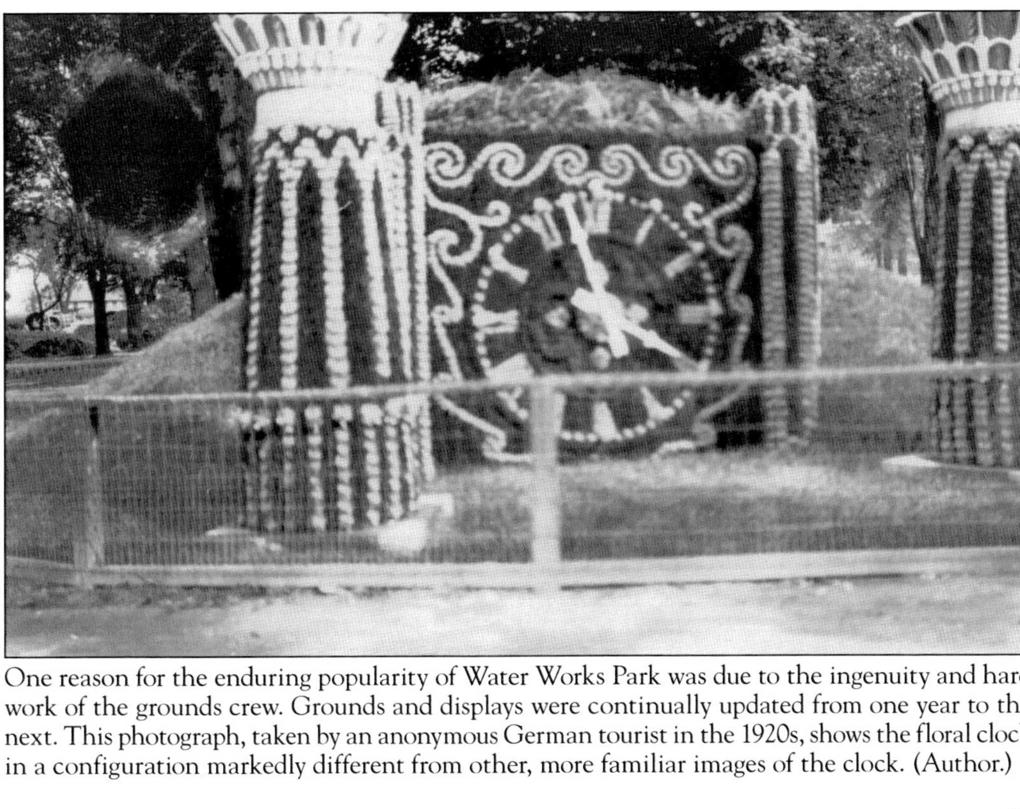

One reason for the enduring popularity of Water Works Park was due to the ingenuity and hard work of the grounds crew. Grounds and displays were continually updated from one year to the next. This photograph, taken by an anonymous German tourist in the 1920s, shows the floral clock in a configuration markedly different from other, more familiar images of the clock. (Author.)

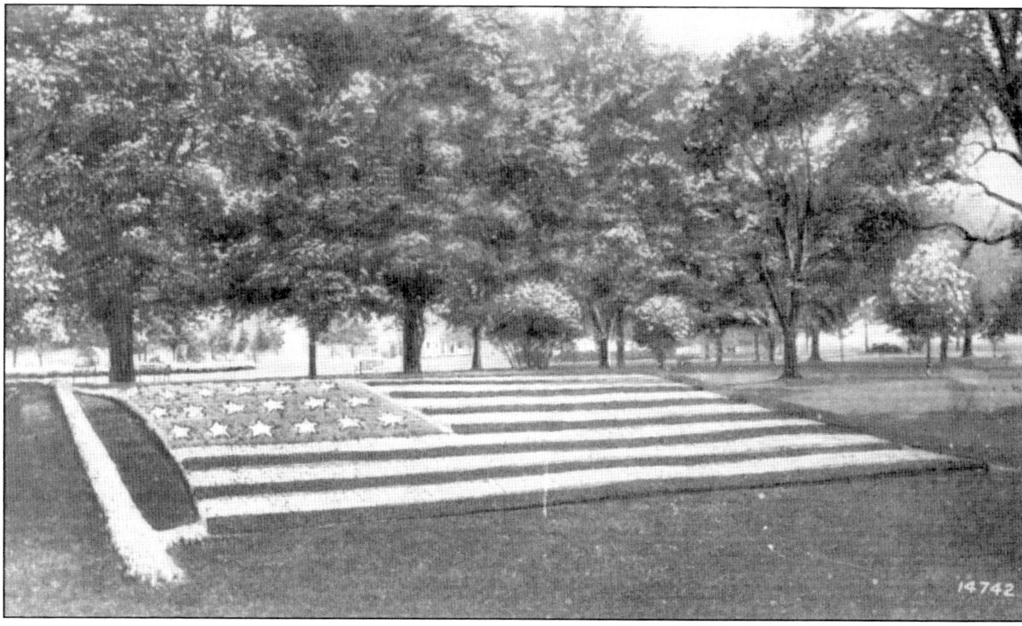

The floral flag was depicted on these two souvenir postcards from around 1910 (above) and around 1900 (below). Skilled groundskeepers recreated the Stars and Stripes from brightly colored blossoms cultivated in the park's two greenhouses. The flag was popular with visitors to Water Works Park for many years. Most visitors to the park never gave much thought to the thousands of gallons of drinking water constantly flowing just beneath their feet. Despite the city's addition of new water treatment plants in 1935, 1956, 1964, and 1973, Water Works Park continues to be a vital source of high-quality drinking water for metropolitan Detroit, and has been for more than 130 years. (Both, author.)

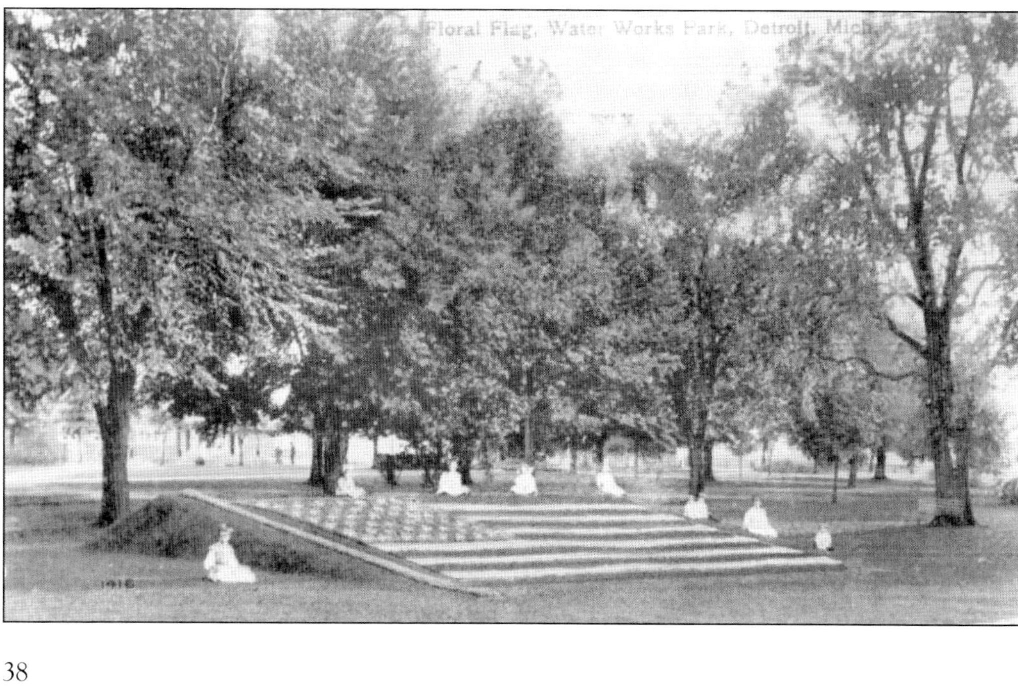

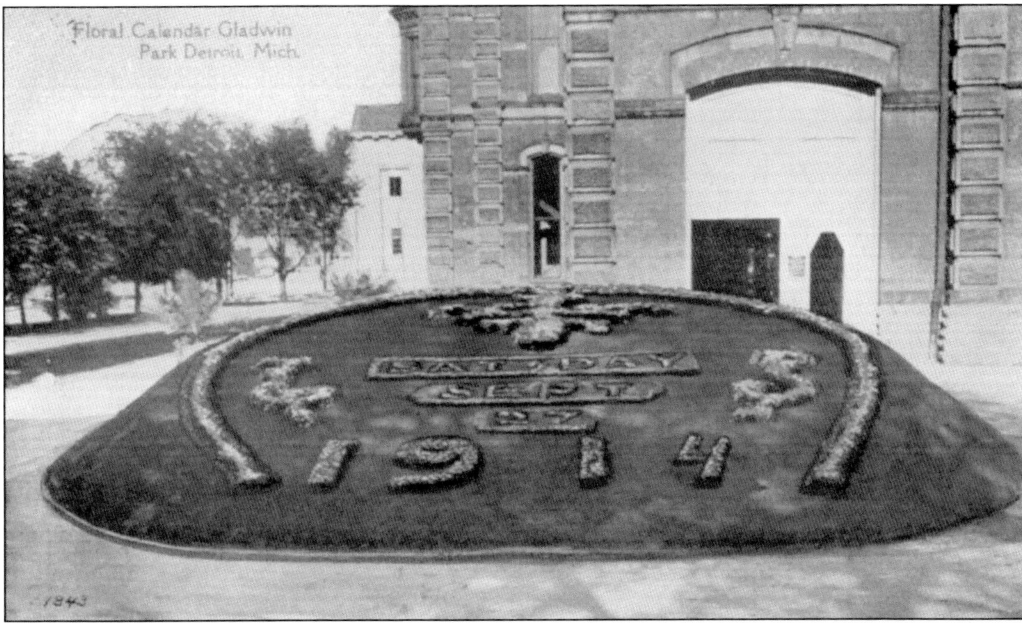

The floral calendar appeared on souvenir postcards as well—including these two from the 1910s. The calendar was located on a small knoll near the original pumping station on the western end of the park, in the vicinity of the standpipe. The ingenious design of the calendar by the park's superintendent of grounds, Elbridge A. Scribner, made it easy for groundskeepers to keep it current by moving around or switching out a few flowers at a time. The multi-talented "Scrib" also worked for a time as a locksmith, operated a small florist shop at 604 East Fort Street, and, with his brother, operated a small hotel known as the Scribner House at 257 Jefferson Avenue. (Both, author.)

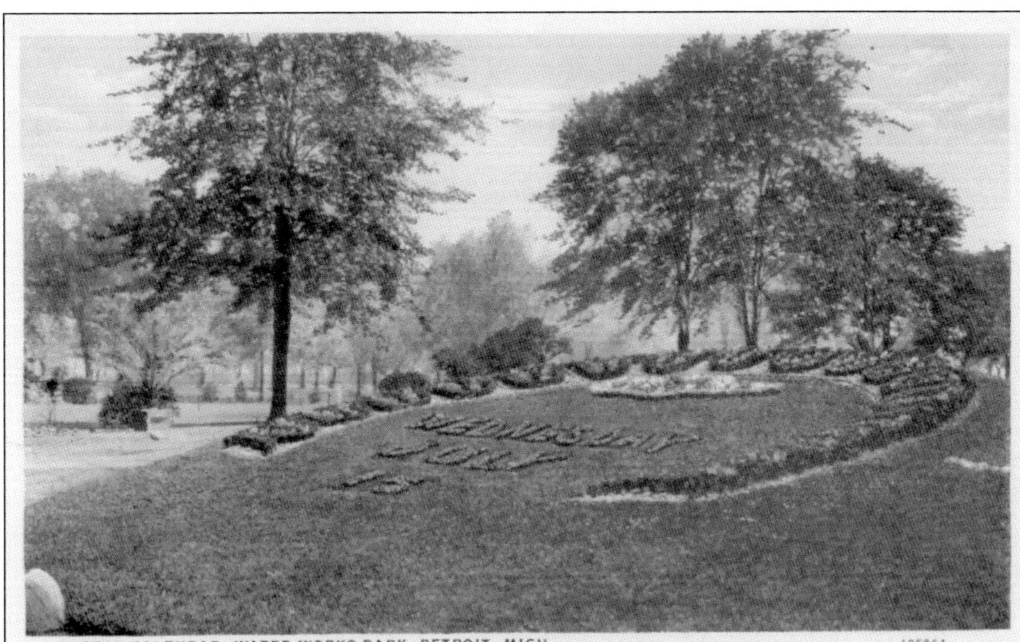

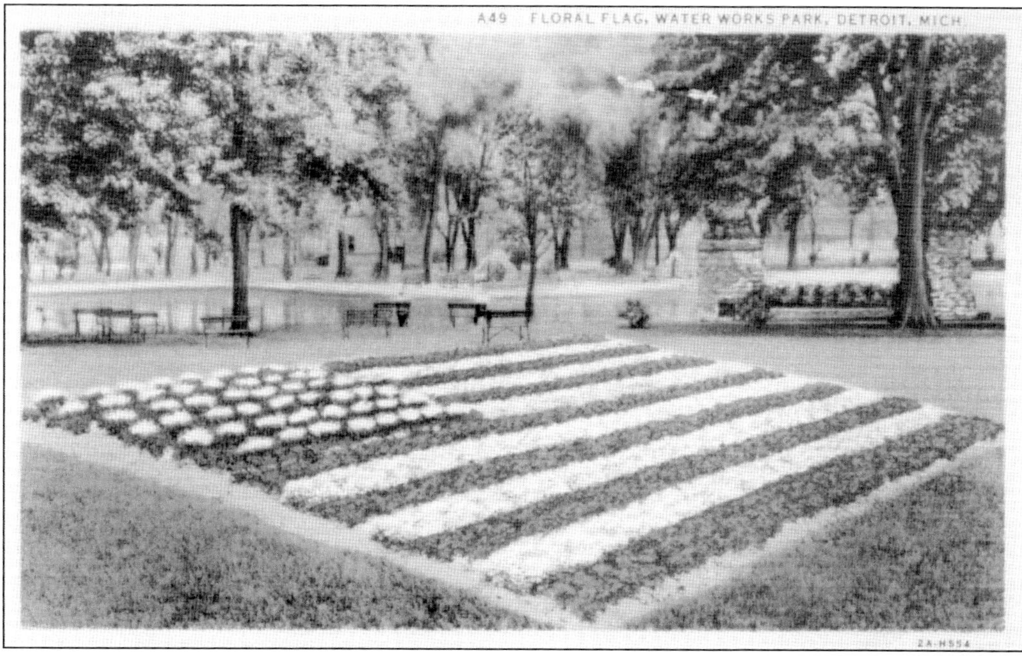

This image of the floral flag from the mid-1920s differs from images of the flag in other years. This was on purpose; crews continually updated the grounds, while maintaining their familiarity. (Author.)

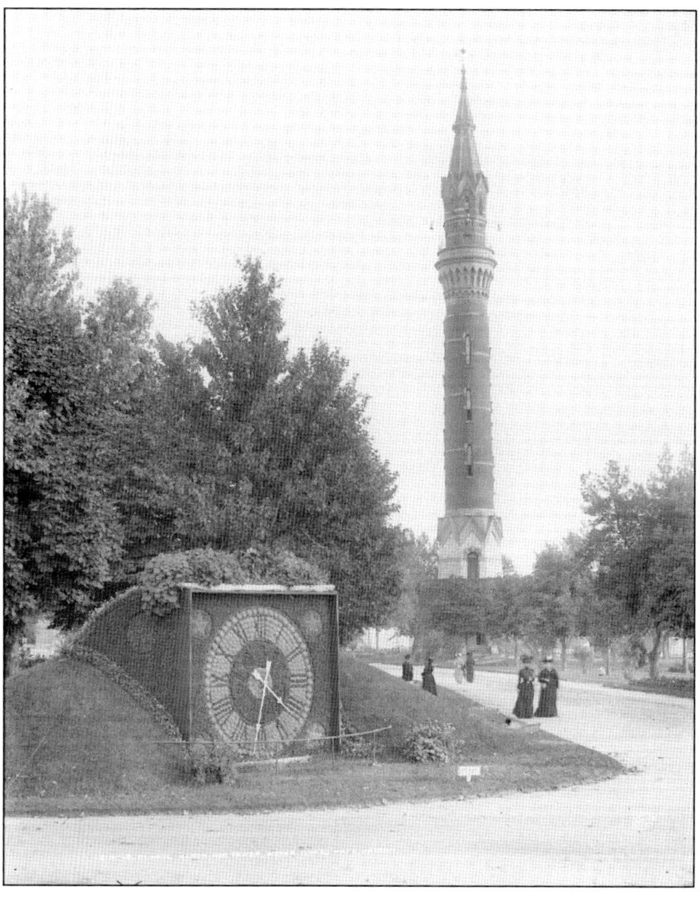

This 1902 photograph shows the clock guarded by a tiny "Keep Off" sign, which was a (likely futile) attempt by groundskeepers to discourage children from hanging from the hands of the clock. (DPC-LOC.)

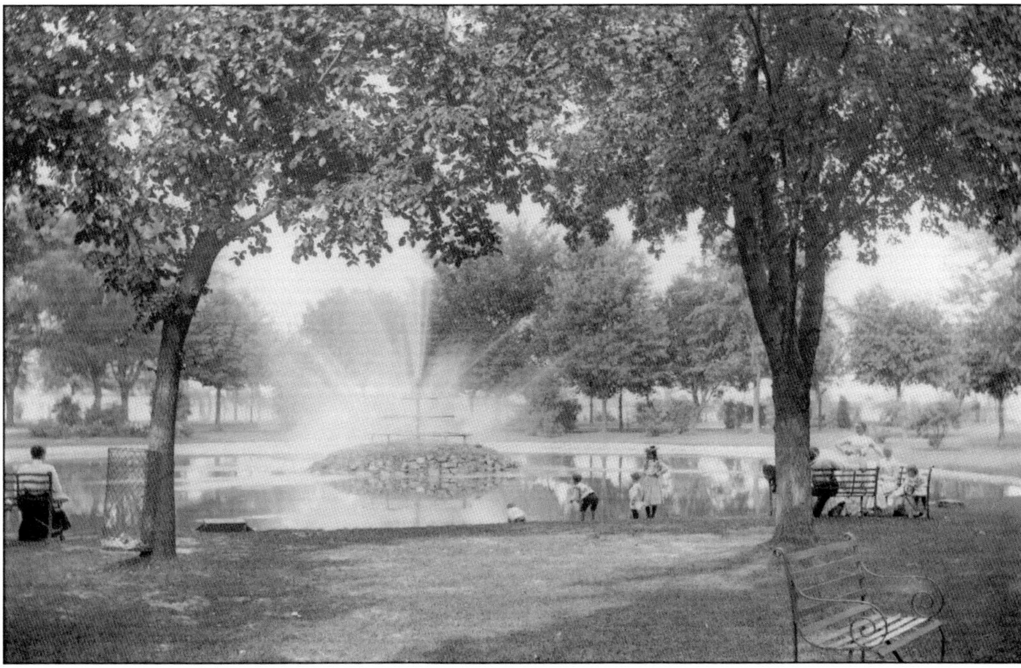

Children play around the lake, or wading pond, around 1900. By 1893, the park contained a man-made waterway with two islands, three bridges, a wading pool, and a winding canal to accommodate rowboats entering from the Detroit River. Swarms of fish were attracted by water that was warmer than that of the river flowing past the park. (Author.)

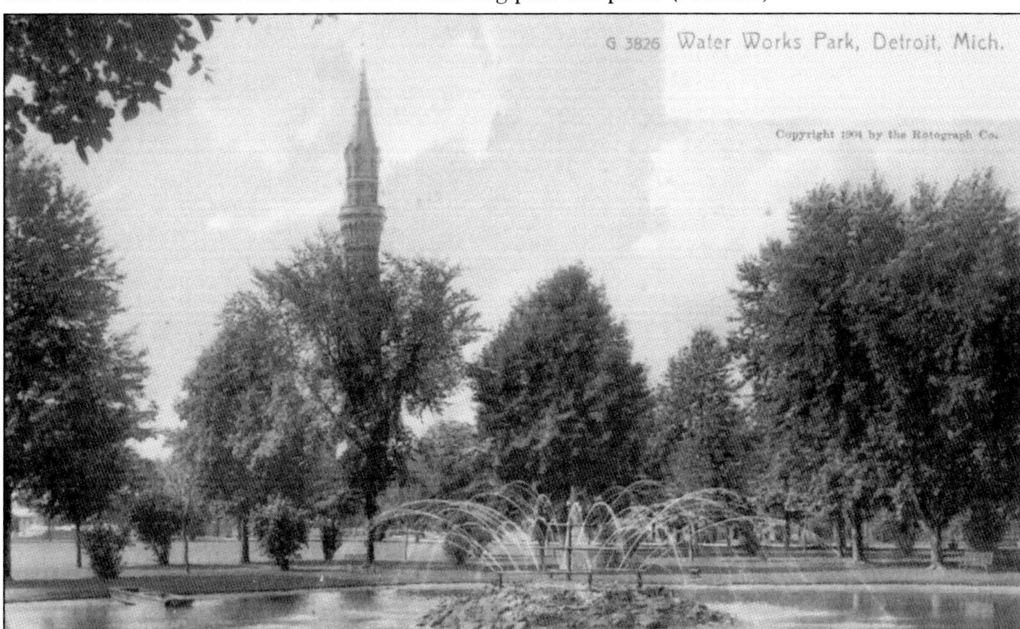

This idealized postcard of the lake and fountain, with the water tower in the distance, is from 1904. The lake was actually a wading pond that underwent a variety of configuration-changes over the years until the remaining version was filled in in 1962. The lake was a popular attraction for people of all ages. (Author.)

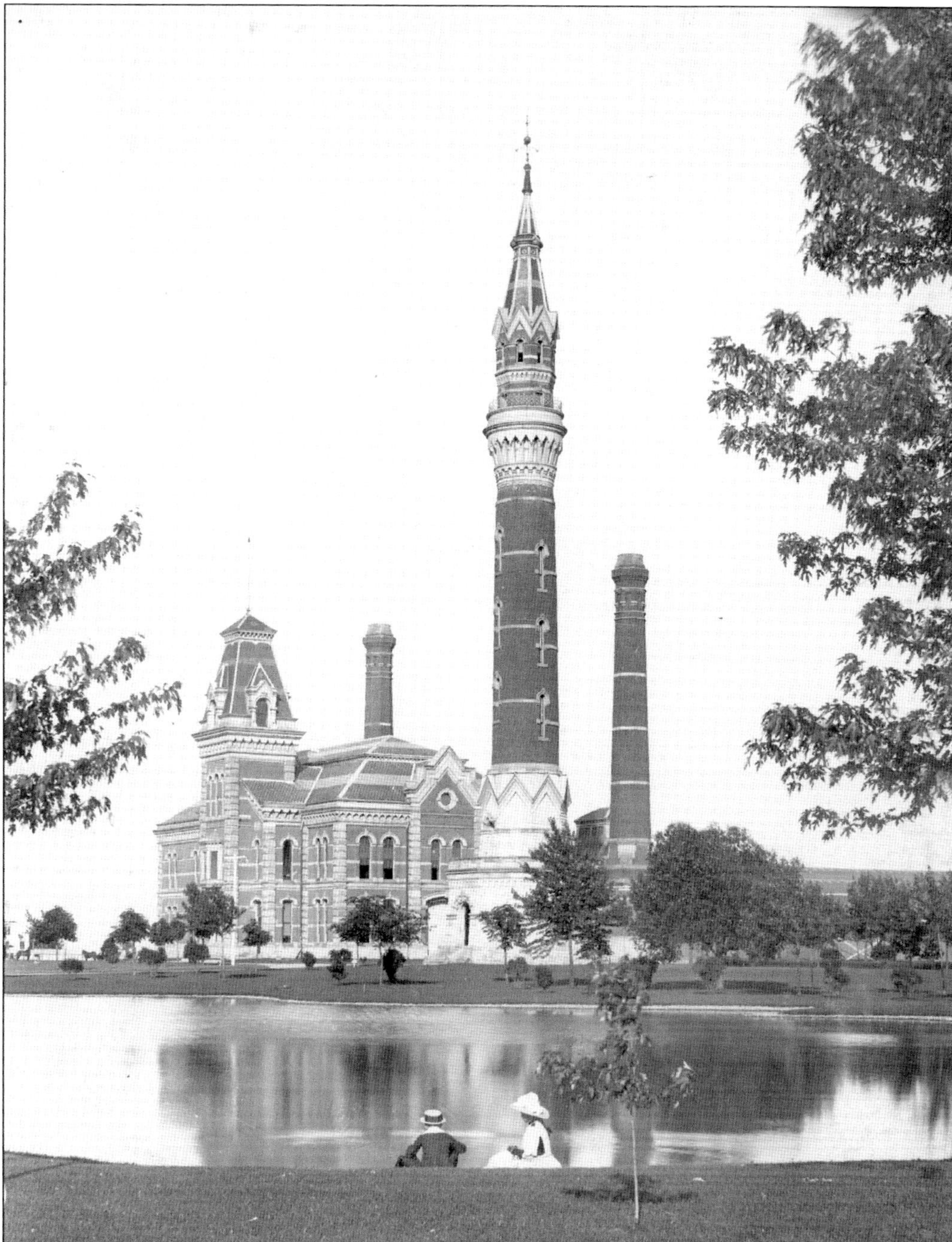

This couple sits by the lake across from the water tower and pumping station. The photograph dates to the late 1870s or early 1880s, as the station is still in its original configuration. The photograph is credited to Detroit architect George D. Mason. (Burton Collection.)

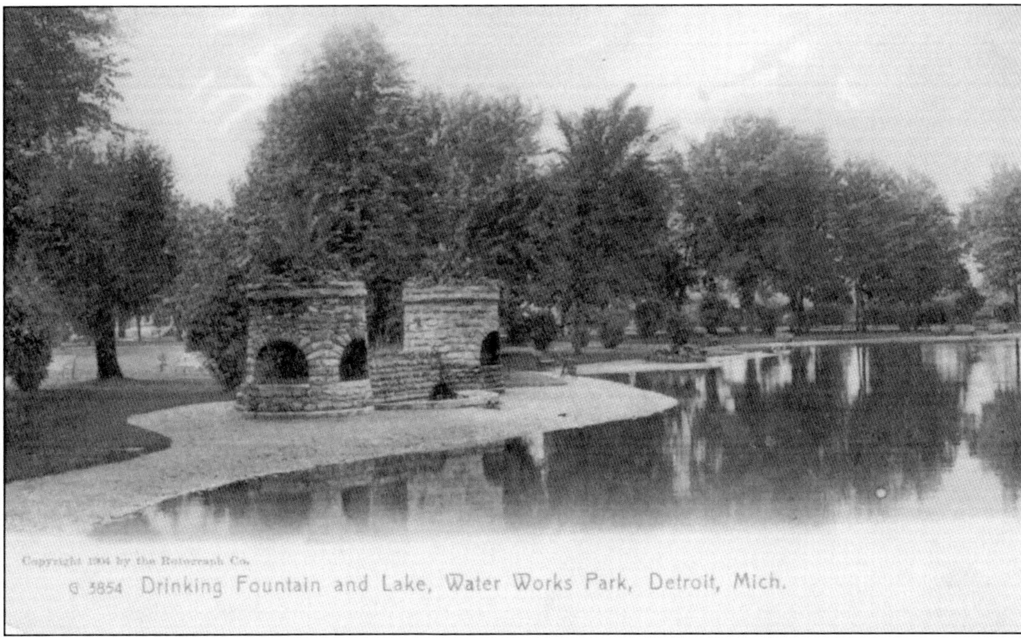

Both of these views of the drinking water fountains, looking east (above) and west (below), are from around 1910. Moses W. Field, an attorney who represented Detroit in the US House of Representatives from 1873 to 1875, is credited with the introduction of public drinking fountains in Detroit. Though the concept of public access to water predates Fields by several centuries, it was his petition to the members of the Detroit Common Council in May 1871 that resulted in the first public fountains on city streets. The city spent a total of $255.24 to install 16 fountains in June and July of that year. (Both, author.)

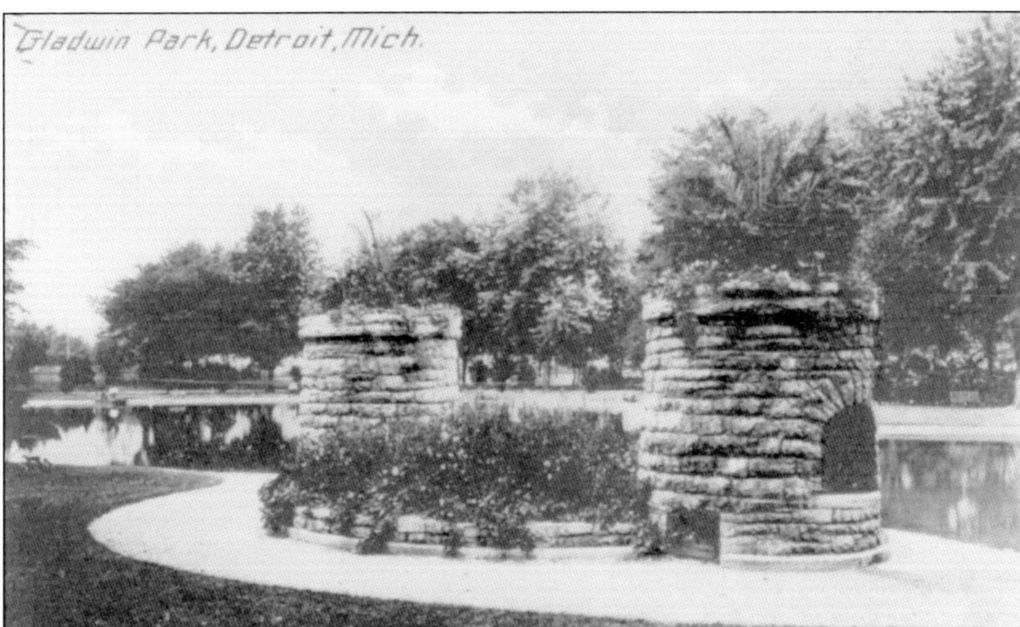

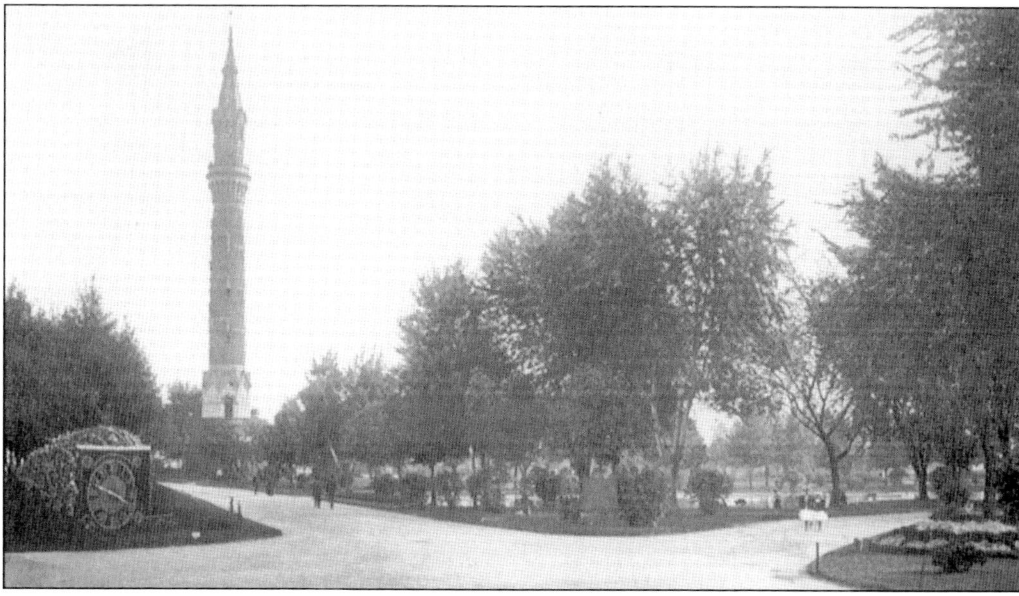

The idea behind Water Works Park as a public gathering space comes from 19th-century city-planning concepts. Large American cities at the time were dirty, tightly packed places. People who lived in them highly prized the airy, clean, open spaces of public parks, and the growing number of parklike cemeteries. Such spaces called for majestic structures, such as the Hurlbut Memorial Gate, to separate the solitude of the grounds from the congestion and noise outside. The highest expression of this concept can be seen in the above image from the 1890s, looking south to the standpipe from the grand entry. The below image, from around 1910, also illustrates the idyllic park setting, as young boys race toy sailboats on the placid surface of the lake. (Both, DPC/LOC.)

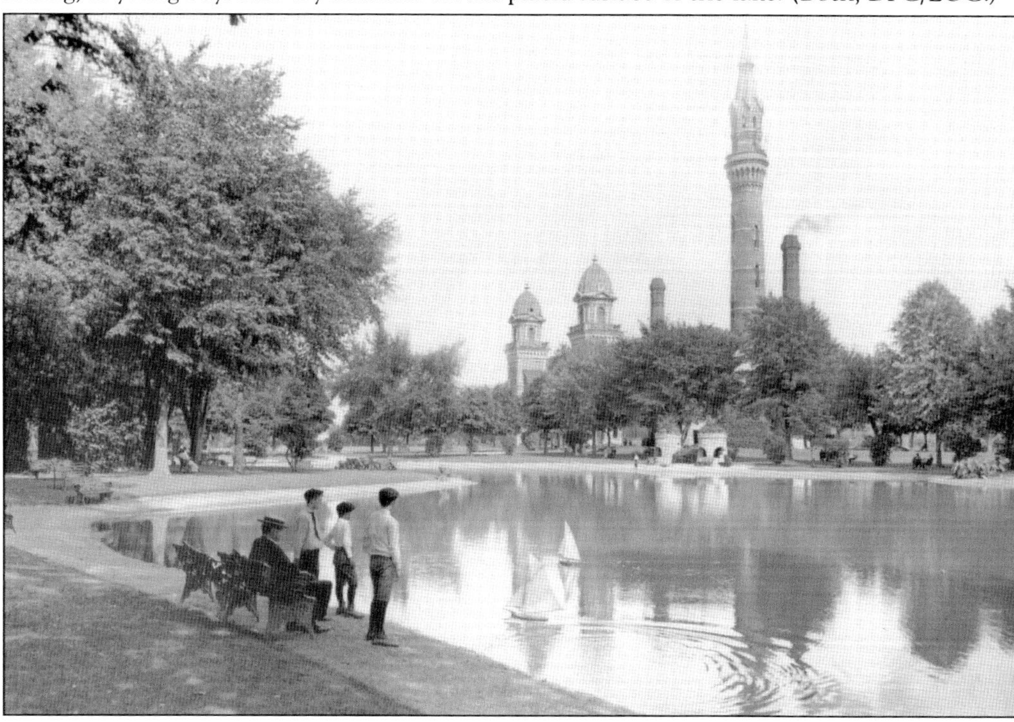

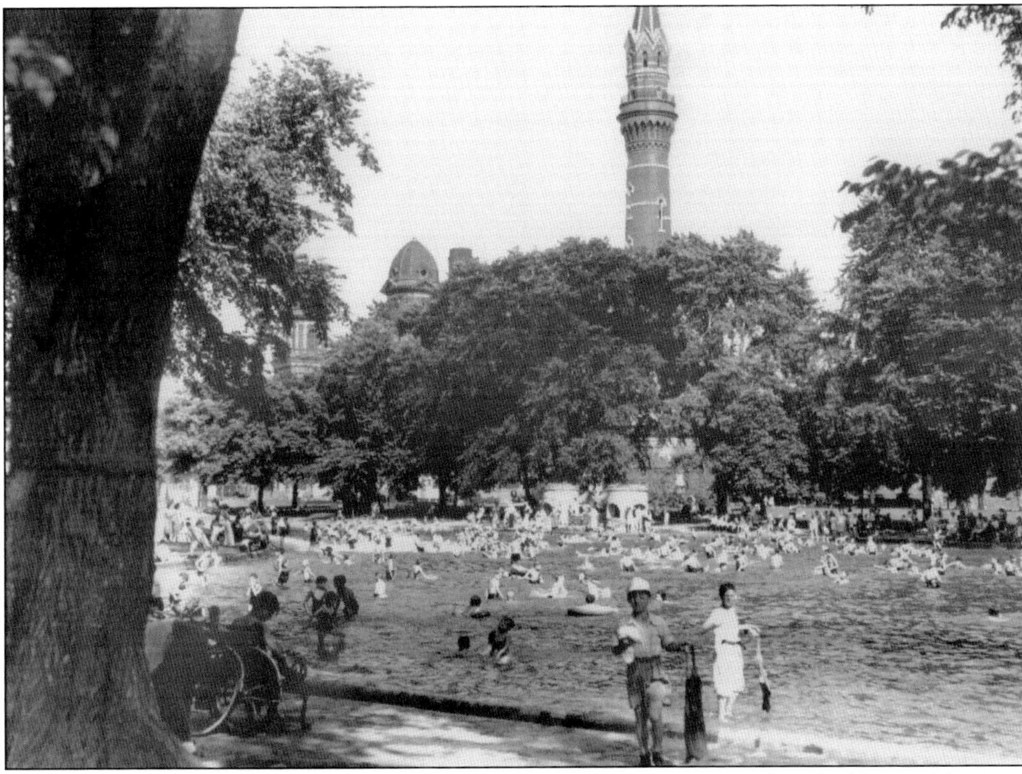

Water Works Park was filled to capacity in June 1936 as Detroiters sought relief from a blistering summer-long heat wave. The temperature hit a near record of 104.4 degrees on July 8 and did not drop below 100 degrees until July 14. The heat claimed the lives of 122 Detroiters, including Pat Brooks, the stepfather of boxer Joe Louis. (*Detroit News*.)

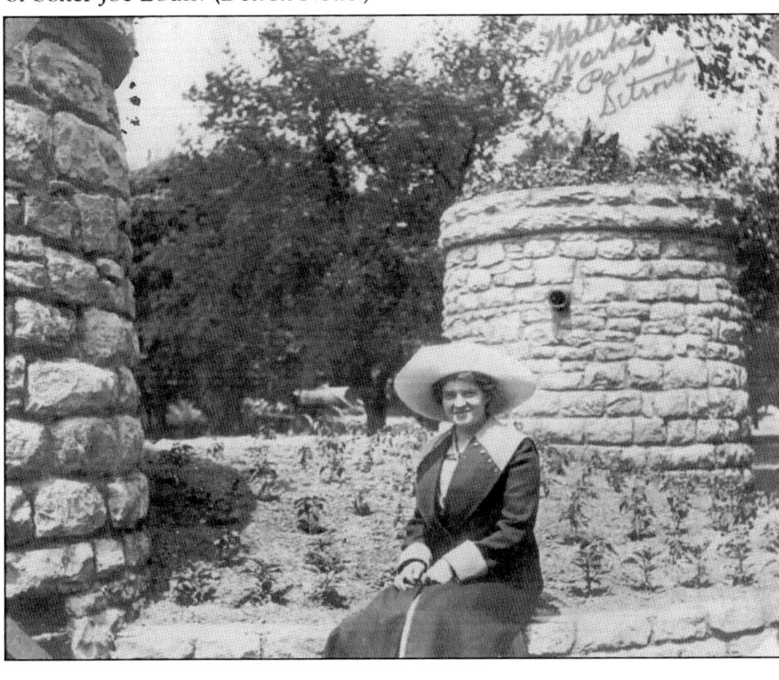

The anonymous woman in this early-20th-century photograph takes a break among the drinking fountains on the lake. Inclusion of the word *Detroit* in the handwritten legend suggests the woman was one of many thousands of people who visited Detroit and Water Works Park from other states and foreign locales. (Author.)

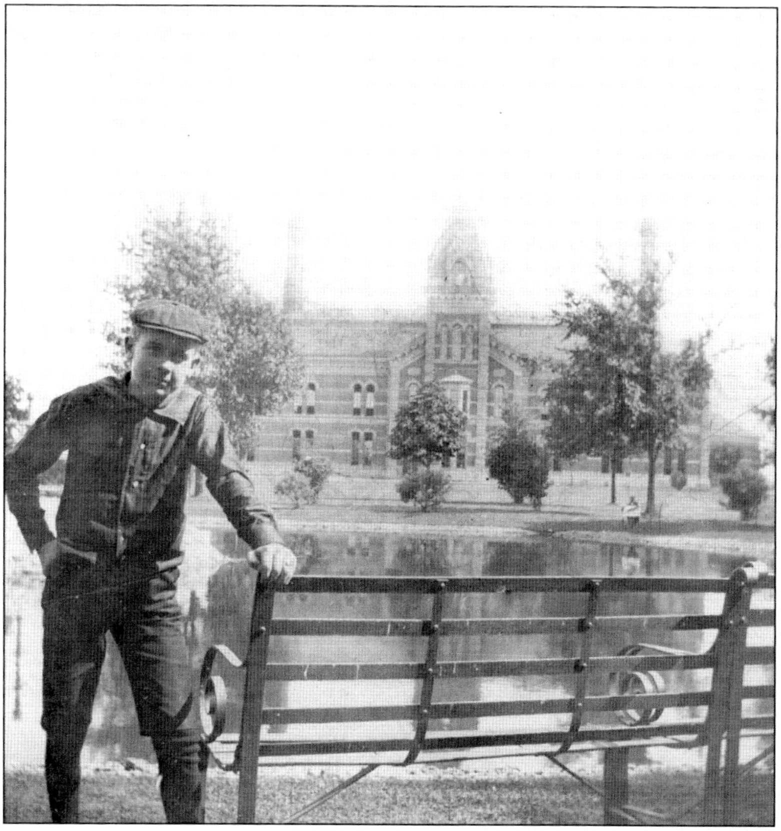

This candid shot of a young boy by a park bench near the lake dates to sometime after 1886, when the station was expanded to make room for a third steam-driven pumping engine. (Burton Collection.)

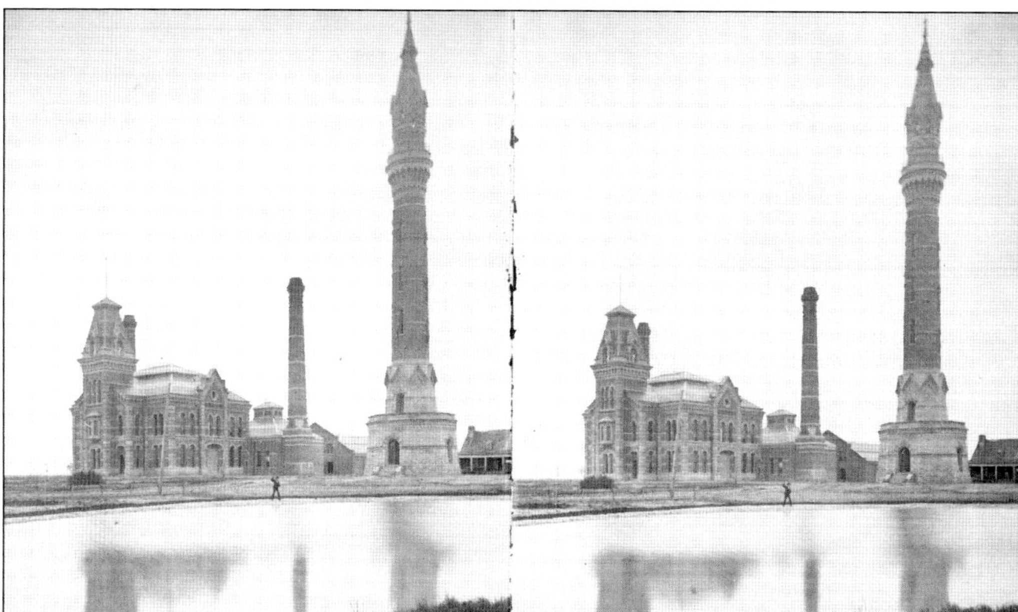

This is a stereoscopic view of the pumping station and water tower in 1881, the year the second steam-driven pumping engine was installed. At this time, the station had only been in operation for two years. (Burton Collection.)

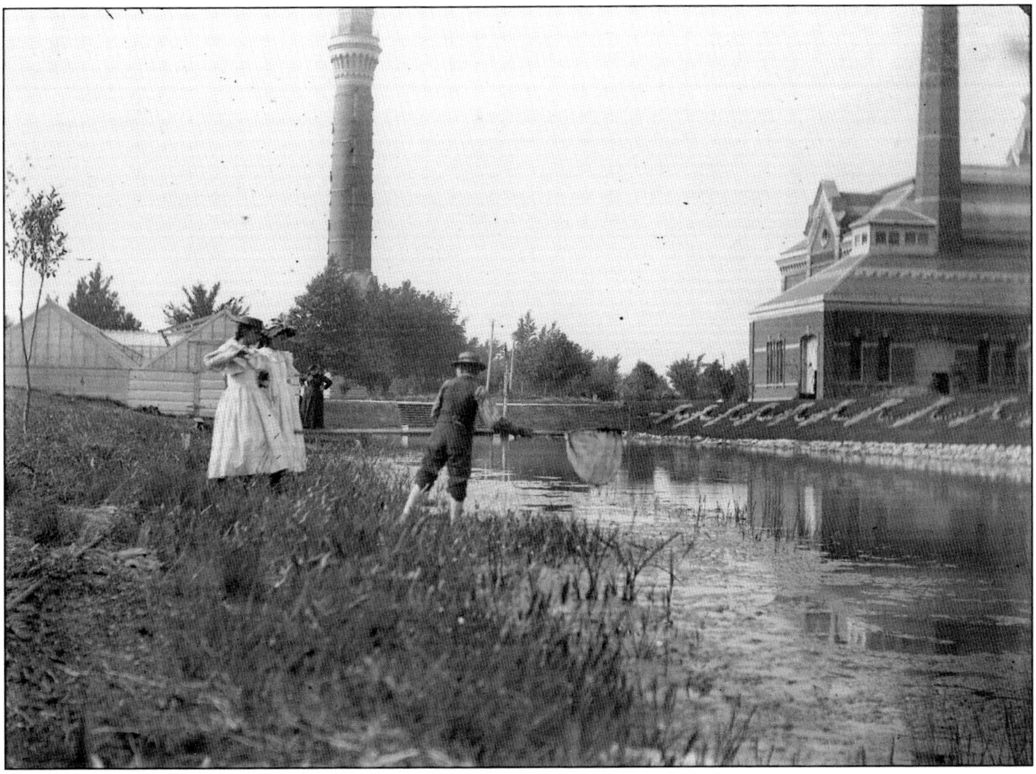

A young boy has an audience as he fishes with a long-handled net in the lake. The greenhouse is to the left behind the two women. This image was captured prior to the construction of the pavilion and probably dates from the early 1880s. (Burton Collection.)

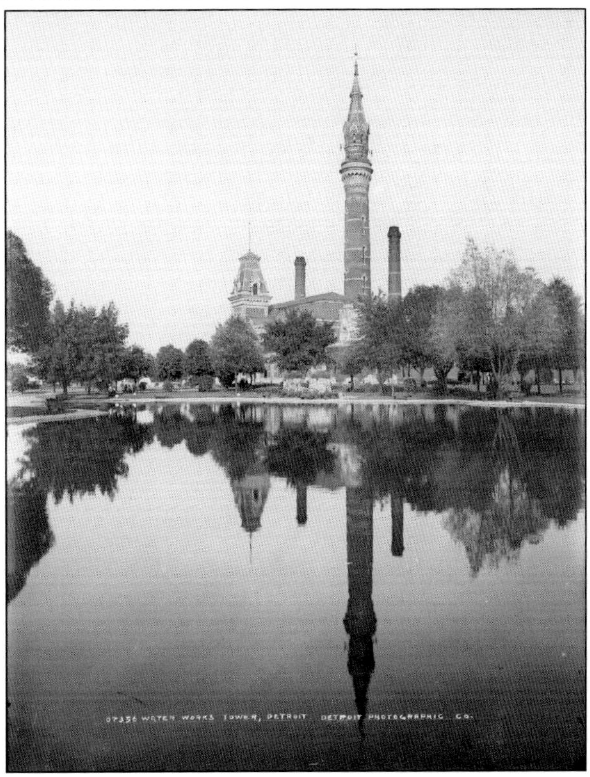

A mirror image of the original pumping station, water tower, and surrounding trees is reflected on the surface of the still water of the lake around 1900. (DPC/LOC.)

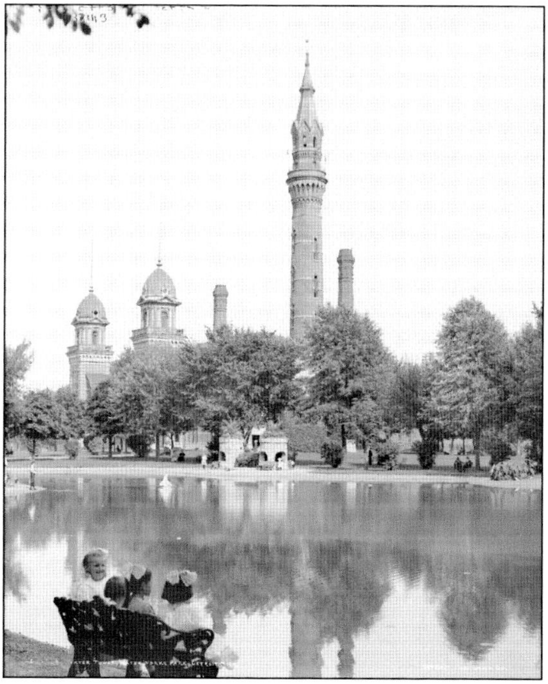

These two classic family-friendly images of the park date from around 1910. In the left image, four young girls enjoy the view from a park bench in the shade on an idyllic summer day. Perhaps three of those same girls have been captured a few minutes before or after in the hand-tinted image below. By 1900, Water Works Park was second in popularity only to Belle Isle among the city's many beautiful public parks. (Left, DPC/LOC; below, author.)

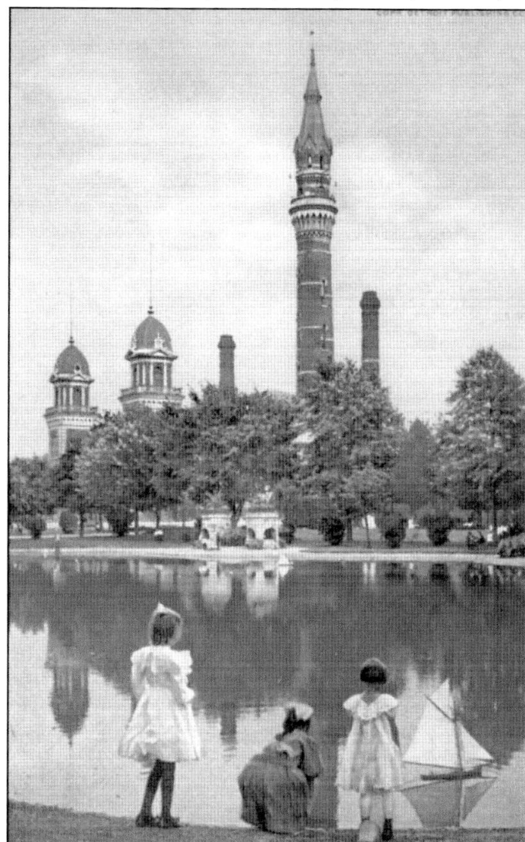

This idealized view shows young boys in knickers skipping stones across the shallow lake in about 1890. Water Works Park was popular among Detroiters of all ages and persuasions, and it was one of the places a person had to take family and friends visiting from out of town. (DWSD.)

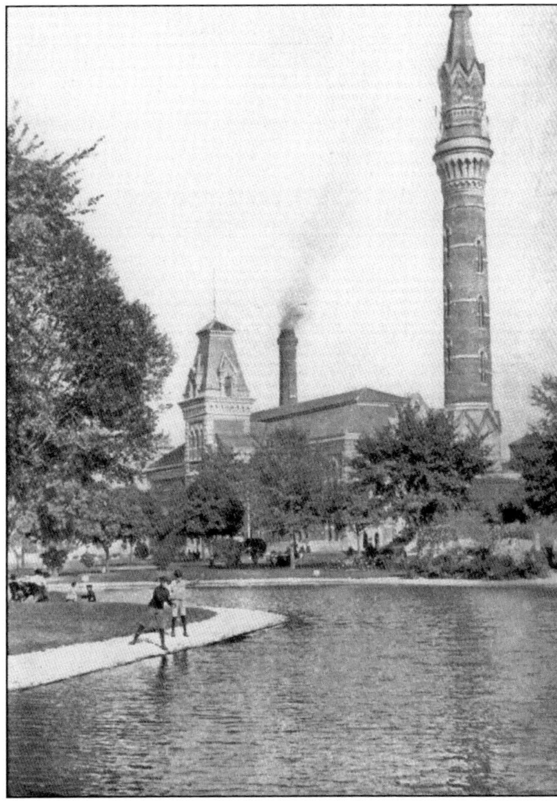

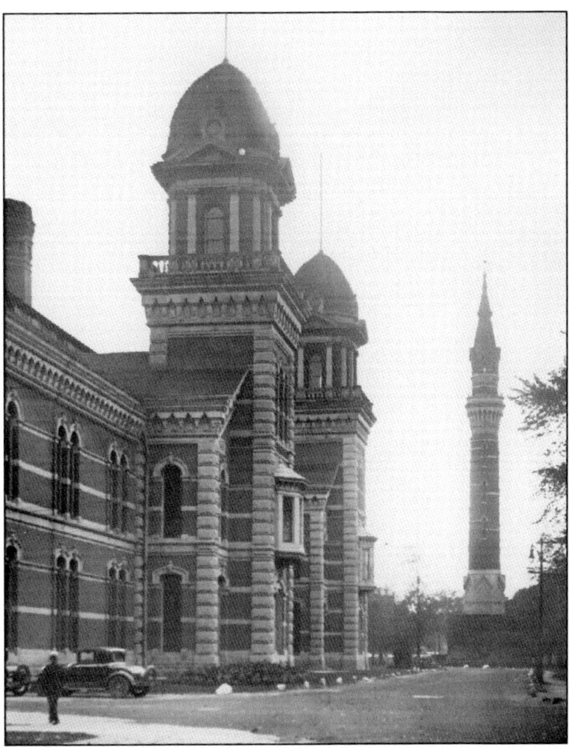

This 1932 image looks west toward the iconic water tower from the front of Pumping Station No. 1. At this time, it was still the exclusive supplier of drinking water to Detroit and its neighboring communities. (Author.)

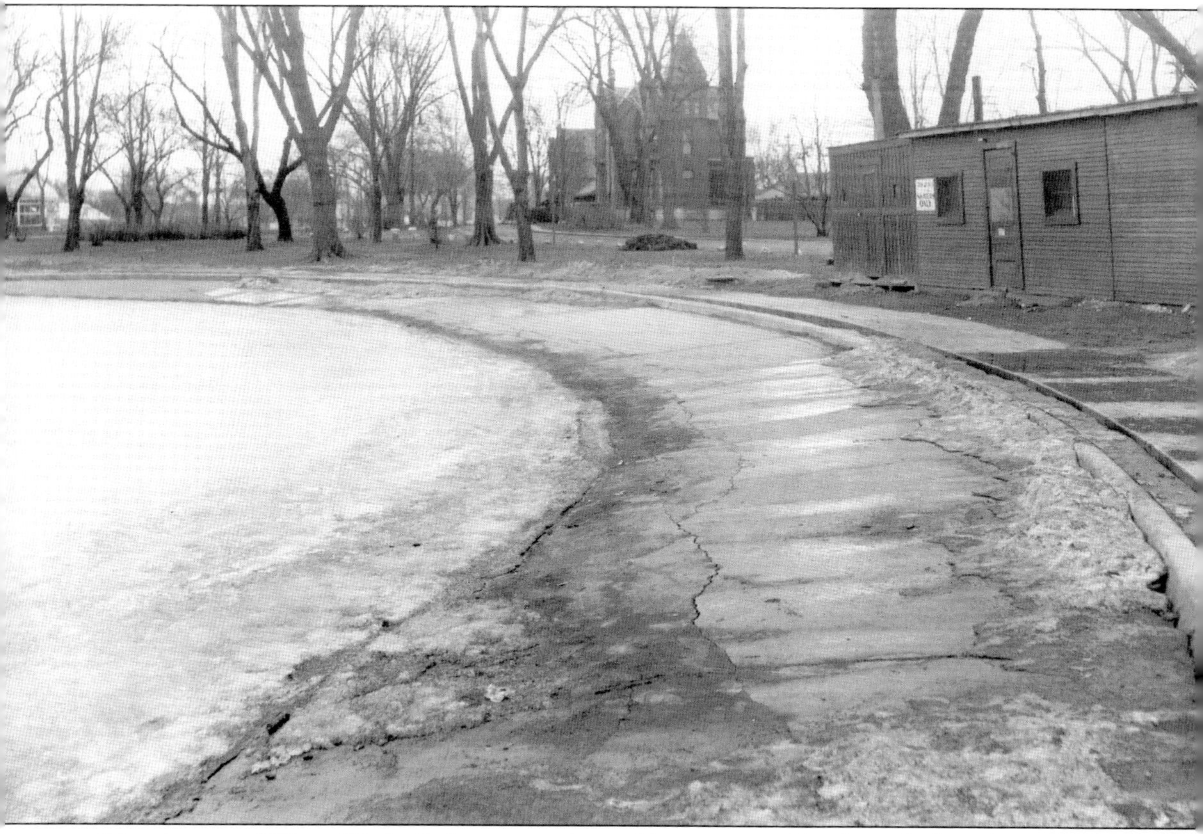

Most images of Water Works Park are snapshots of lazy, summerlike days. Such is the collective memory of those who went there in their youth. But this view of an ice-covered lake, and the sign "This Place is for Skaters Only" from February 1931, proves that public access to the park was year-round. (DWSD.)

This image of the water tower from around 1905 includes some of the park's original French pear trees, the Twelve Apostles. Only a handful of French pear trees—once a signature species of the Detroit River region—remain; most of them are in Grosse Pointe, Monroe, and Windsor. (DPC/LOC.)

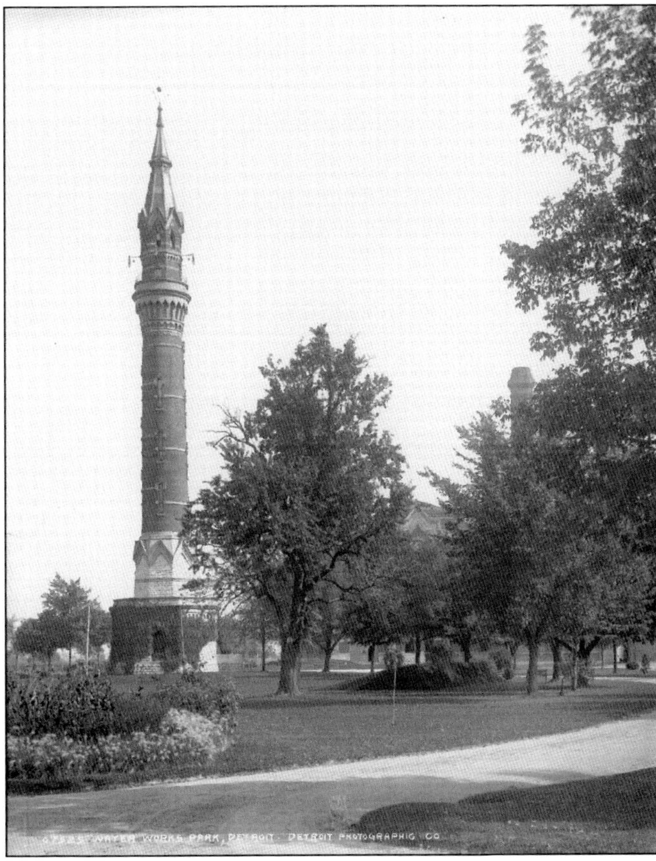

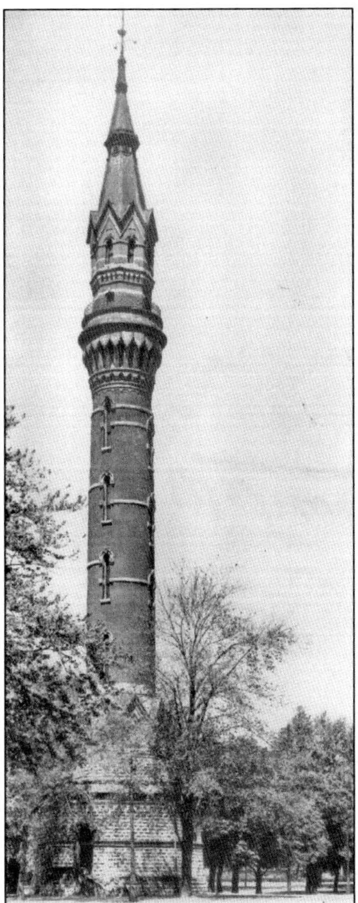

Making its debut in Water Works Park not long after Custer's defeat on the banks of Little Big Horn in 1876, the water tower, at 185 feet tall, was the tallest structure in the western hemisphere at the time. Visitors reached the observation platform near the top by climbing the 202 iron steps ringing the open-ended, 124-foot-tall standpipe within the tower. (Author.)

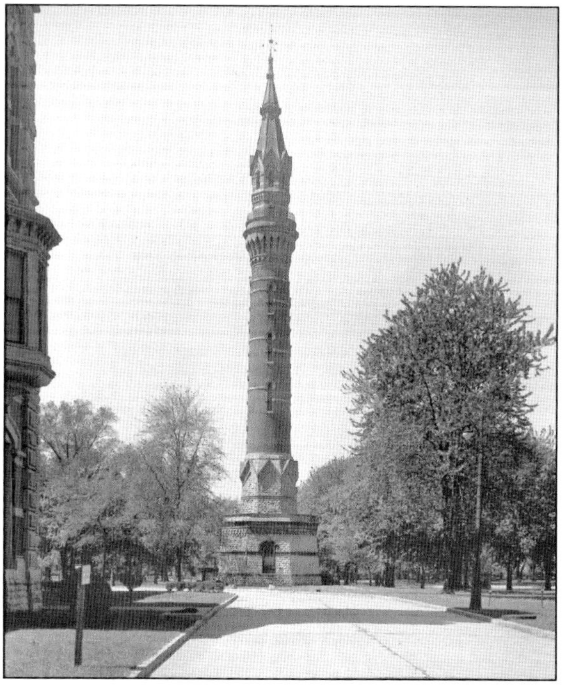

The open-ended standpipe was enclosed within the Victorian-and-minaret-inspired brick water tower designed by Detroit architect Joseph E. Sparks. At its introduction in 1876, the tower, at 185 feet, was the tallest structure in Detroit and one of the world's tallest at a time when few structures exceeded three or four stories. The design submitted by Sparks, a relatively obscure Detroit architect, was chosen over those submitted by more prominent Detroit architects such as George D. Mason. The tower has been described as "an architectural exclamation point." Former Detroit mayor and city council president John C. Lodge called it "the Empire State Building of its day." (Left, DWSD; below, author.)

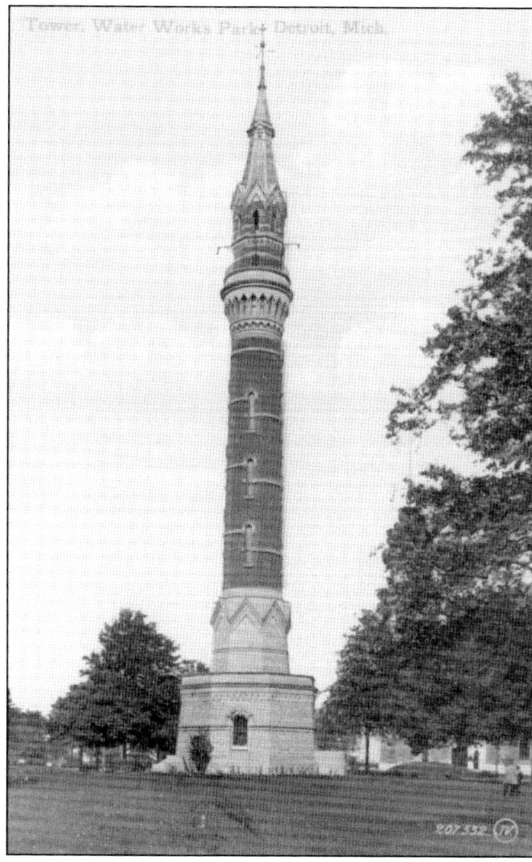

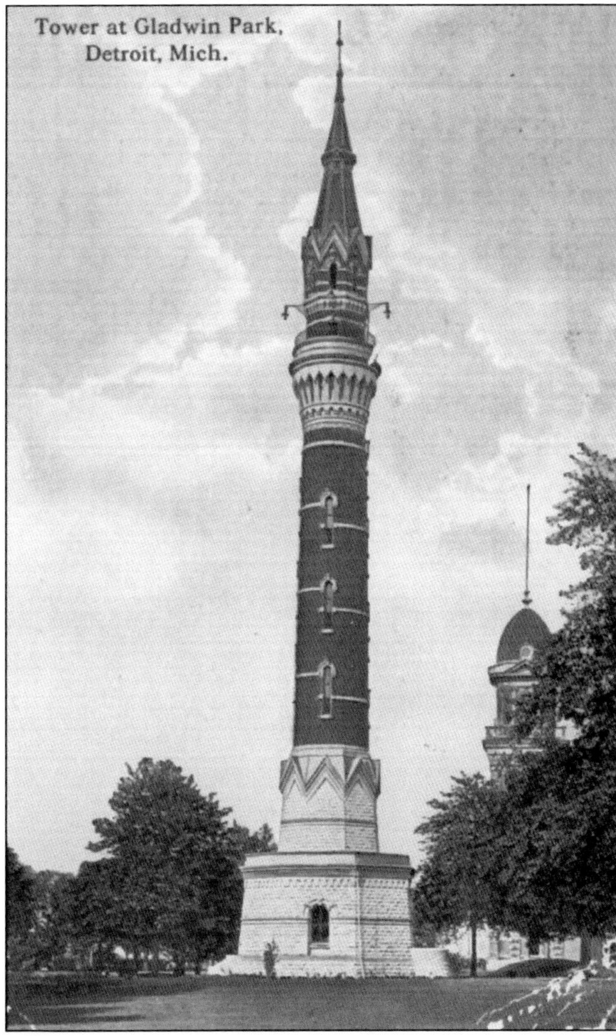

The iconic water tower started life as a functioning part of Detroit's water distribution system. The actual standpipe enclosed in the tower was designed to protect the pipes in the water distribution system from damage caused by high-pressure surges from the big steam-driven pumps in the original pumping station. By 1895, however, the need for pressure equalization by the tower was eliminated by new technology that allowed output pressures to be controlled by the pumps themselves. The structure continued to serve for another 50 years as an observation tower and popular tourist attraction for Detroit. (Right, author; below, DWSD.)

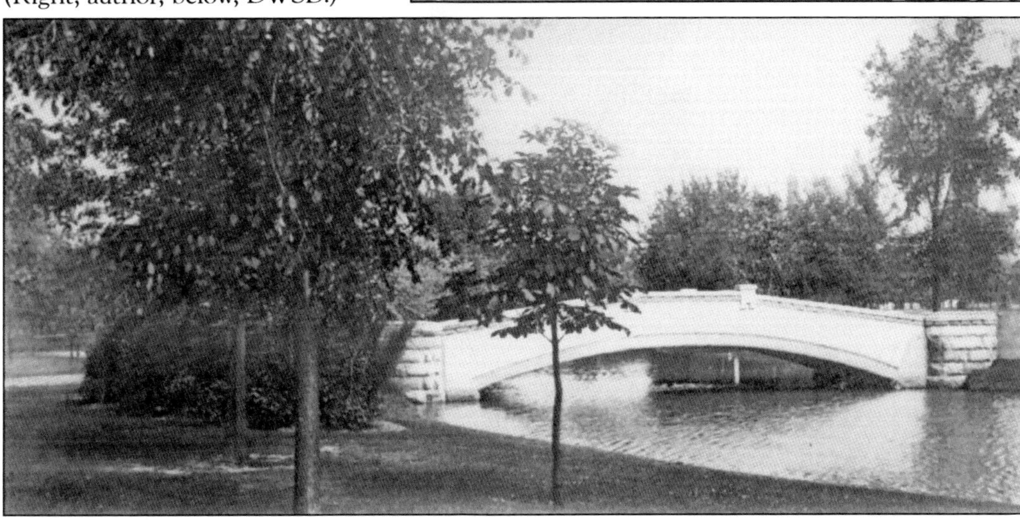

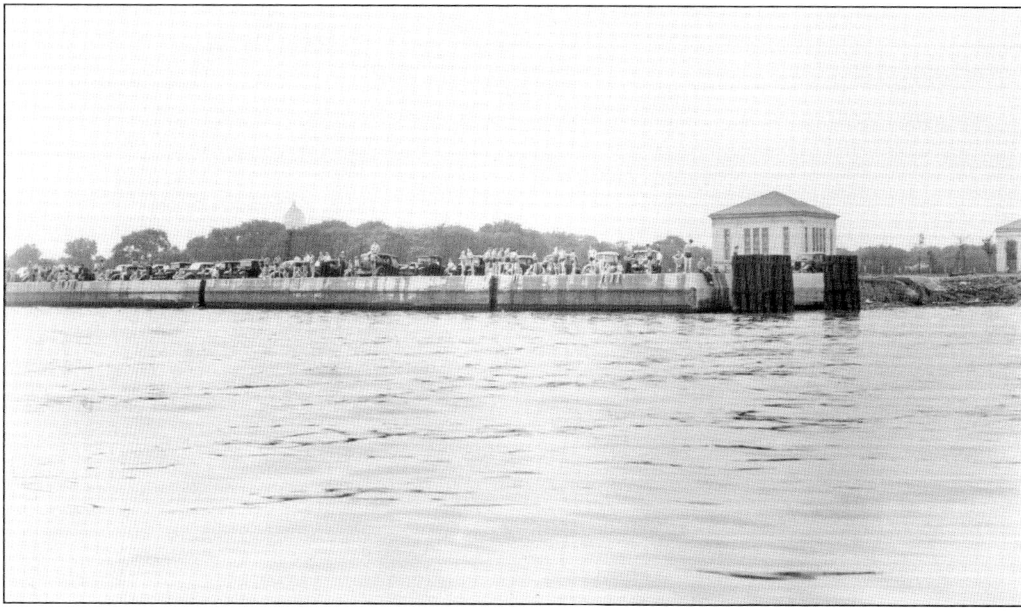

Shown here is the dock on the Detroit River at the foot of Water Works Park in 1939. This location has been a favorite of fishermen over the years and is also used by fans during Gold Cup–class hydroplane racing events on the river each year in July. The structure to the right is the water department's pre-chlorination building, an integral component of the water treatment process. (Author.)

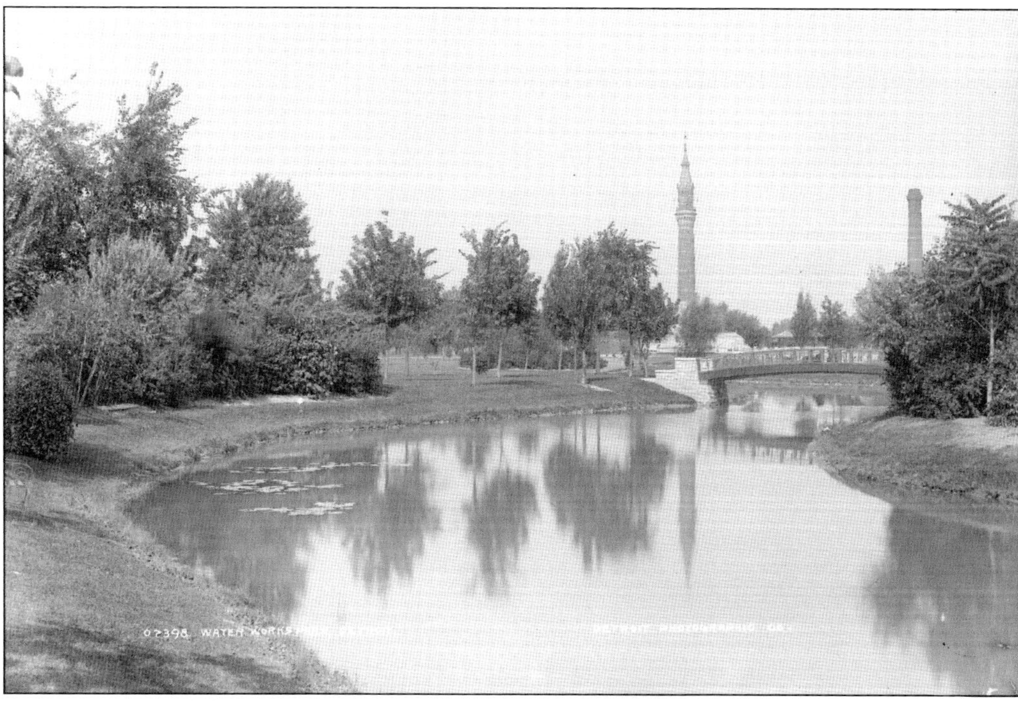

An 1890s view of the winding canal looks northward toward the water tower in Water Works Park. One of two, this canal allowed entry to the park from the Detroit River by visitors in canoes and other types of small craft. (DPC/LOC.)

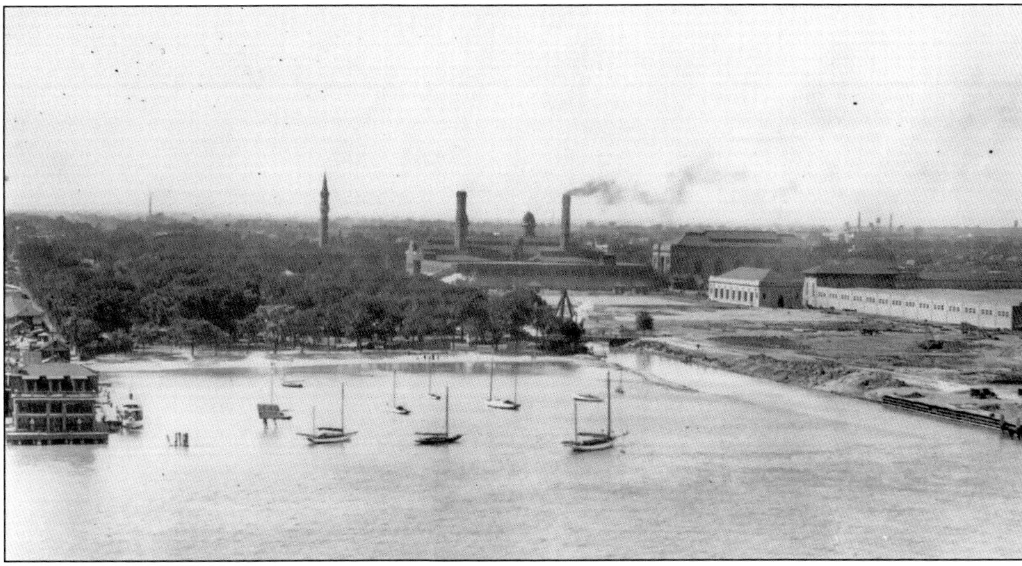

Water Works Park is shown in this aerial view from above the Detroit River on May 25, 1925. The area of the river in the foreground was used as a seaplane-mooring basin for commercial aviation prior to the construction of Detroit City and Metro airports. Visible are the water tower, the world's largest water-treatment plant, and Pumping Stations One and Two. (Author.)

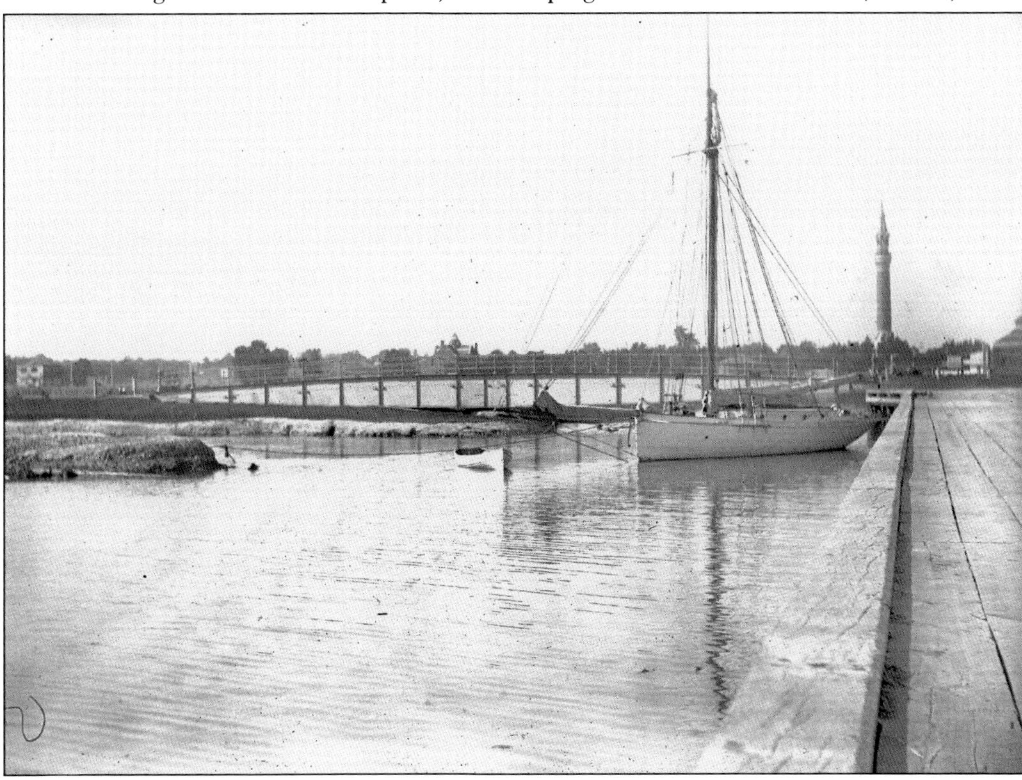

This view from the dock on the Detroit River shows a small sailboat and rowboat moored at the mouth of the winding canal in Water Works Park. In the distance are the water tower and pumping station. (Burton Collection.)

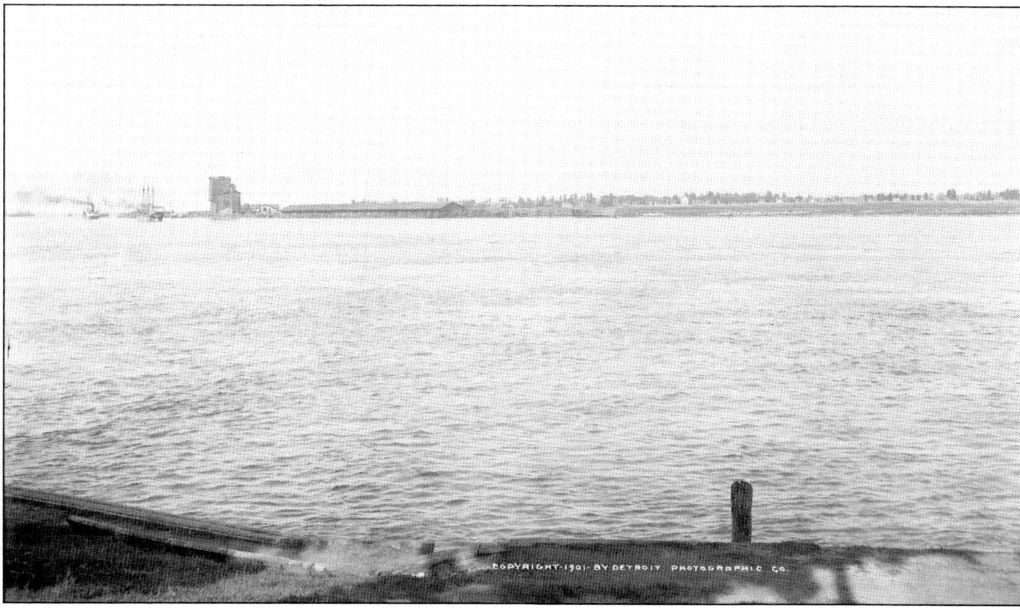

This view of the Detroit River looks upriver toward Lake St. Clair from the foot of Water Works Park in about 1901. Most of the smoke trails appear to originate from freighters. It is highly likely, however, that one or more of the smoke trails are from the stacks of popular excursion steamers that routinely passed by the park on their way to the St. Clair Flats. (DPC/LOC.)

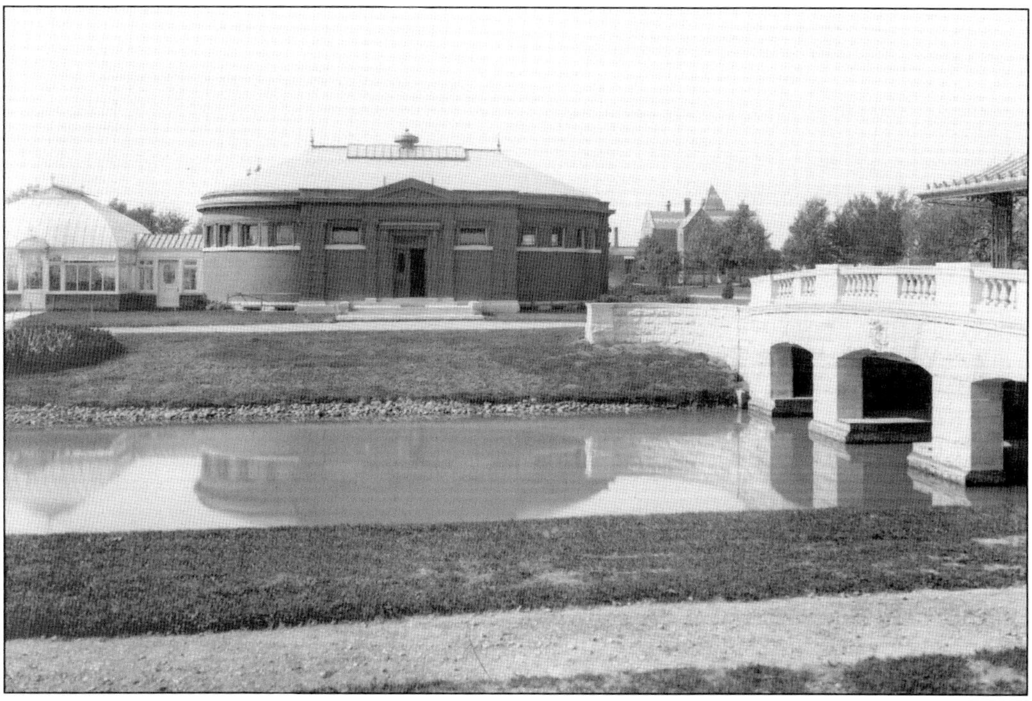

The Hurlbut Library is seen from across the coaling canal in Water Works Park around 1900. A carving of Neptune, the Roman god of water and the sea, is on the bridge to the right on which the pavilion sits. A greenhouse is to the left. The building in the distance is the residence of the head civil engineer of the Detroit Board of Water Commissioners. (Author.)

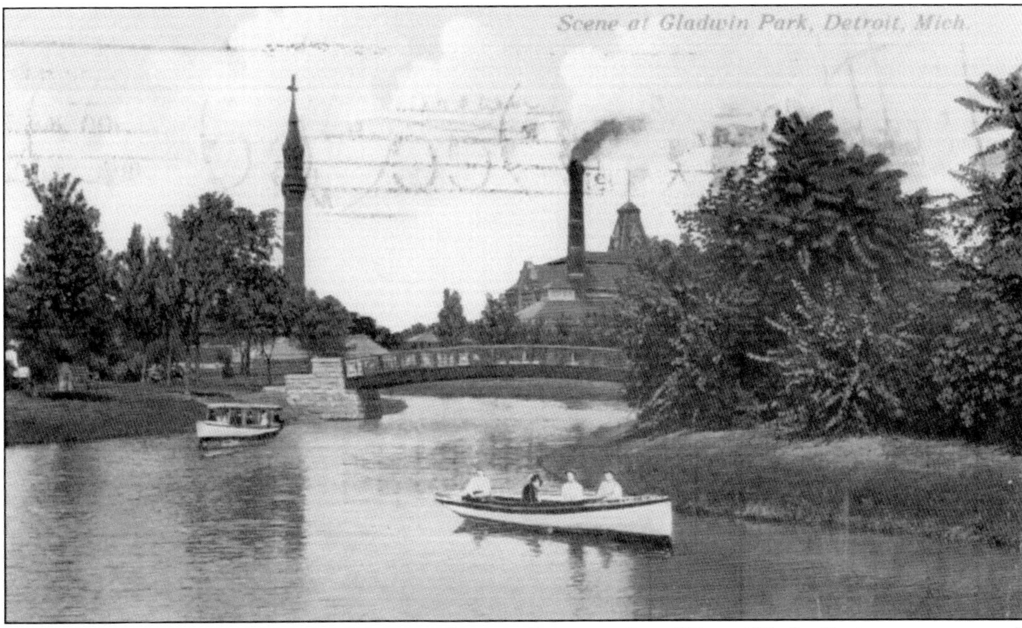

Excursionists navigate their rowboats in the winding canal in Water Works Park near its mouth at the Detroit River around the turn of the 20th century. One of the bridges spanning the canal can be seen beyond the bend. (Author.)

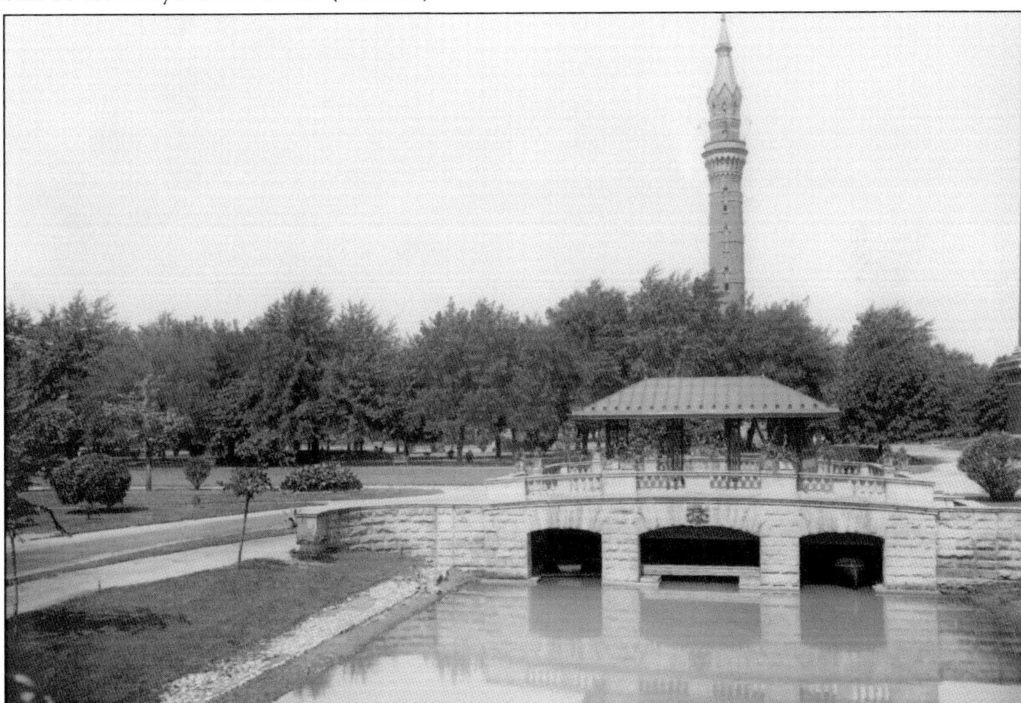

It must have been quite an experience to enter Water Works Park back in the day through the majestic and ornate entrance dedicated to Chauncy Hurlbut. On summer Sundays from 1906, the park would have been alive with music supplied by a live band from the pavilion (shown from behind, looking north) next to the pumping station at the head of the coaling canal. (Author.)

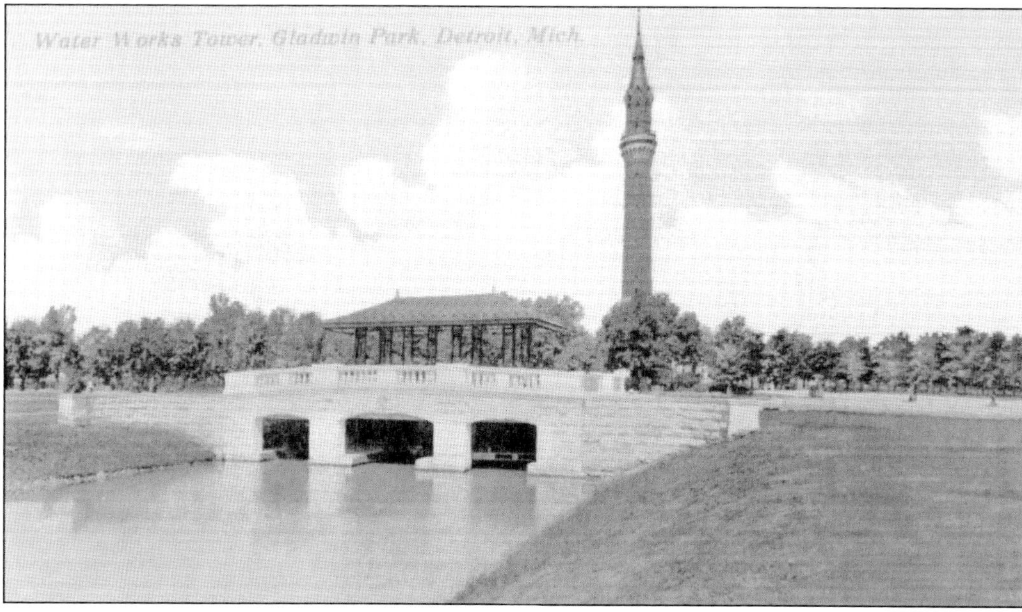

Shown here is an isolated and appealing image of the pavilion and water tower on one of the many popular postcards of Water Works Park. Many of the bands that played from the pavilion on summer Sundays styled themselves after wildly popular "March King" John Philip Sousa and the Marine Corps Band. (Author.)

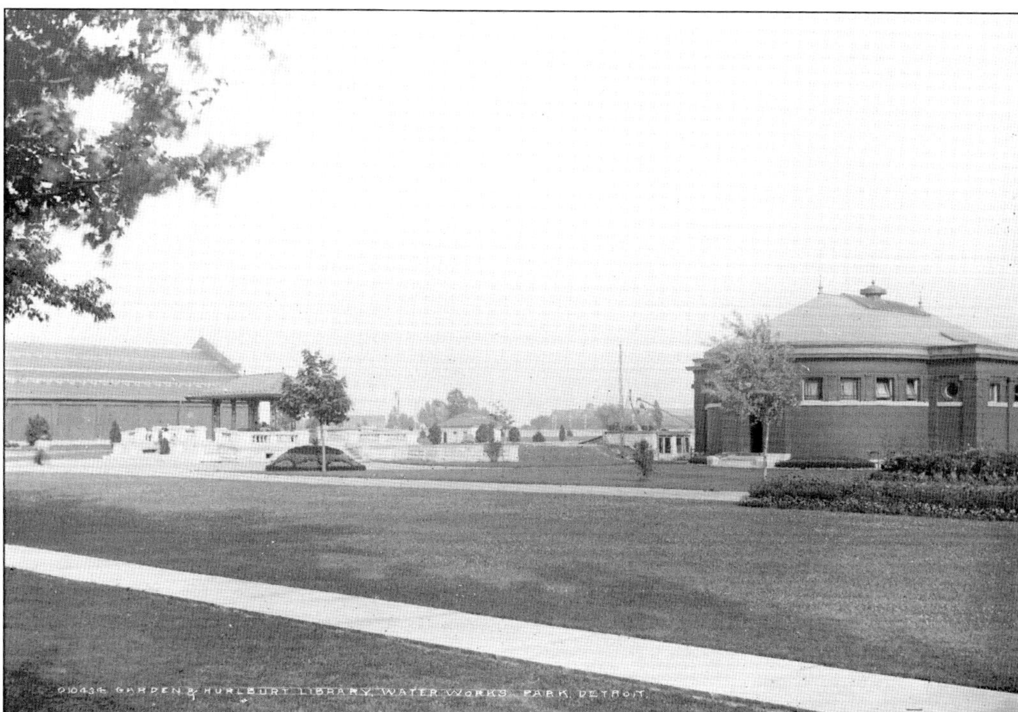

Couples enjoy a day at Water Works Park from the pavilion in the shadow of the Hurlbut Library in the 1890s. The building behind the pavilion is the coal shed. Beginning in 1906, visitors to the park on summer Sundays enjoyed free concerts from bands in the pavilion. (DPC/LOC.)

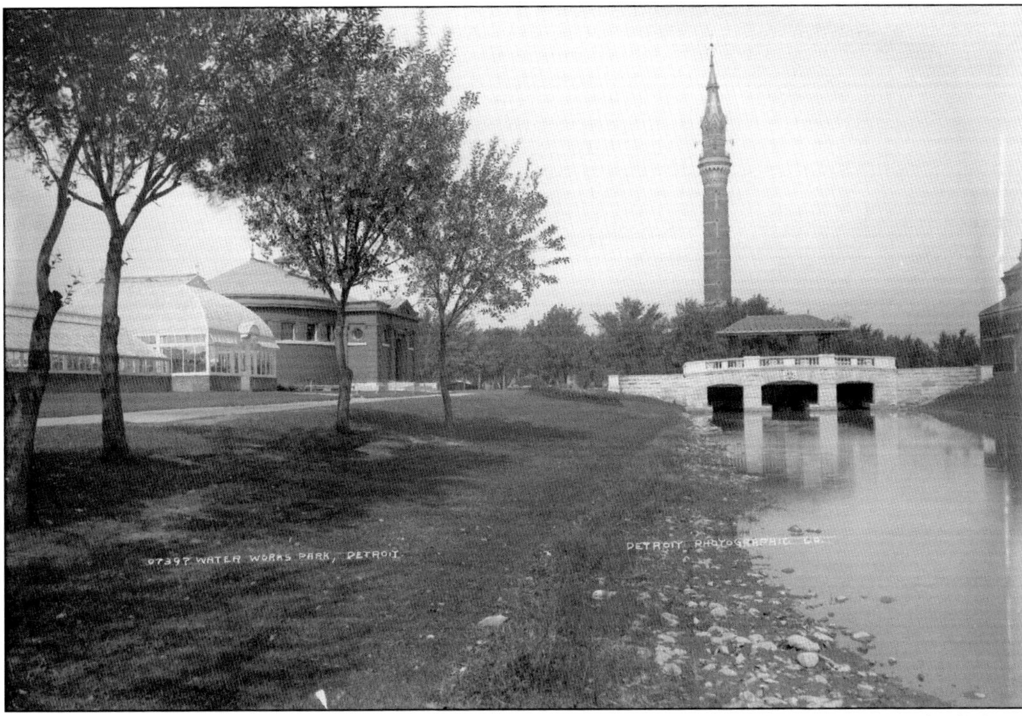

This image is looking north from the canal in Water Works Park in about 1901. At the left are the greenhouses and behind them, the converted oil-storage building used to house the Hurlbut Library. Straight ahead is the pavilion, with the water tower in the distance. The original pumping station is at the extreme right. (DPC/LOC.)

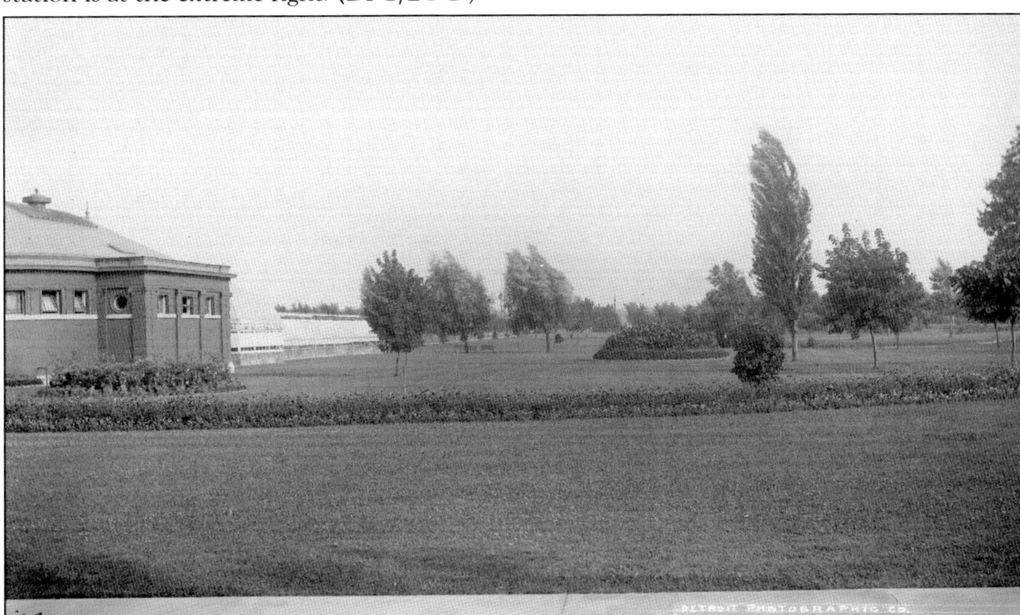

The grounds west of the Hurlbut Library and greenhouses at Water Works Park are shown in the 1890s. The mast in the distance, beyond the trees, could easily belong to a sailing vessel preparing to moor near the mouth of the winding canal beyond the trees on the right. (DPC/LOC.)

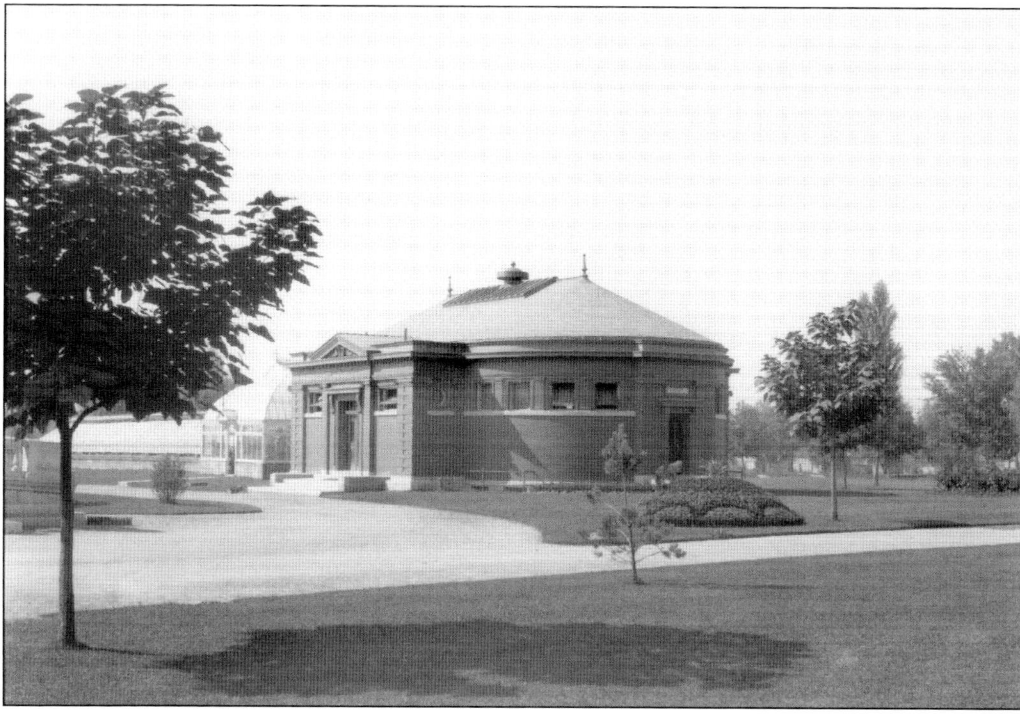

This view of the Hurlbut Library and greenhouses looks southwest on the grounds of Water Works Park. The Hurlbut Library was a former oil-storage building converted in 1897 to house Chauncy Hurlbut's large collection of technical books and periodicals. It was the first satellite branch of the Detroit Public Library system. (DPC/LOC.)

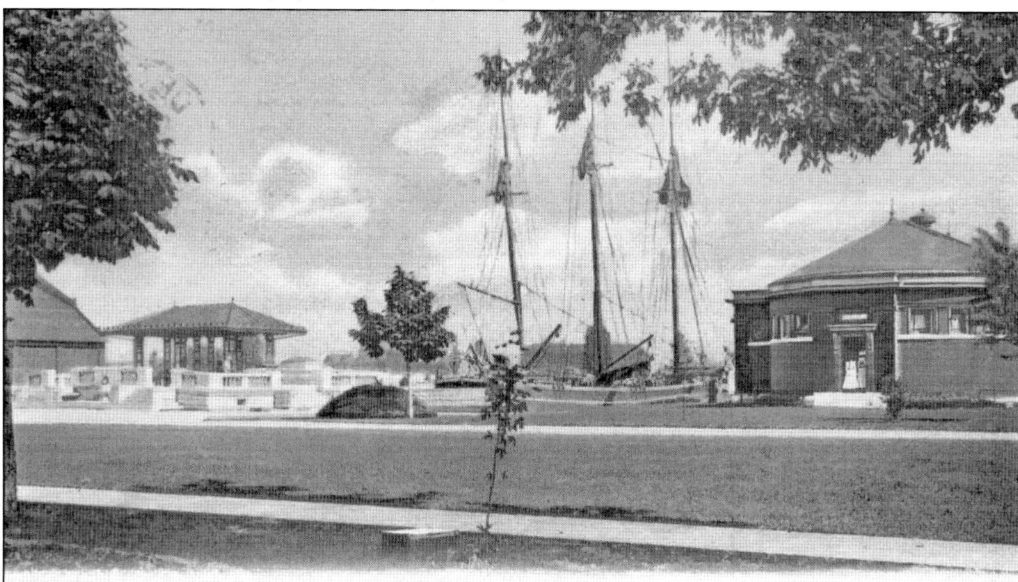

The pavilion (left) and the Hurlbut Library seem to serve as bookends for a three-masted ship at Water Works Park on a hand-tinted postcard from around 1900. The vessel, a coaling ship in the canal, is unloading its valuable cargo to feed the engines in the pumping station just out of sight to the left. (Author.)

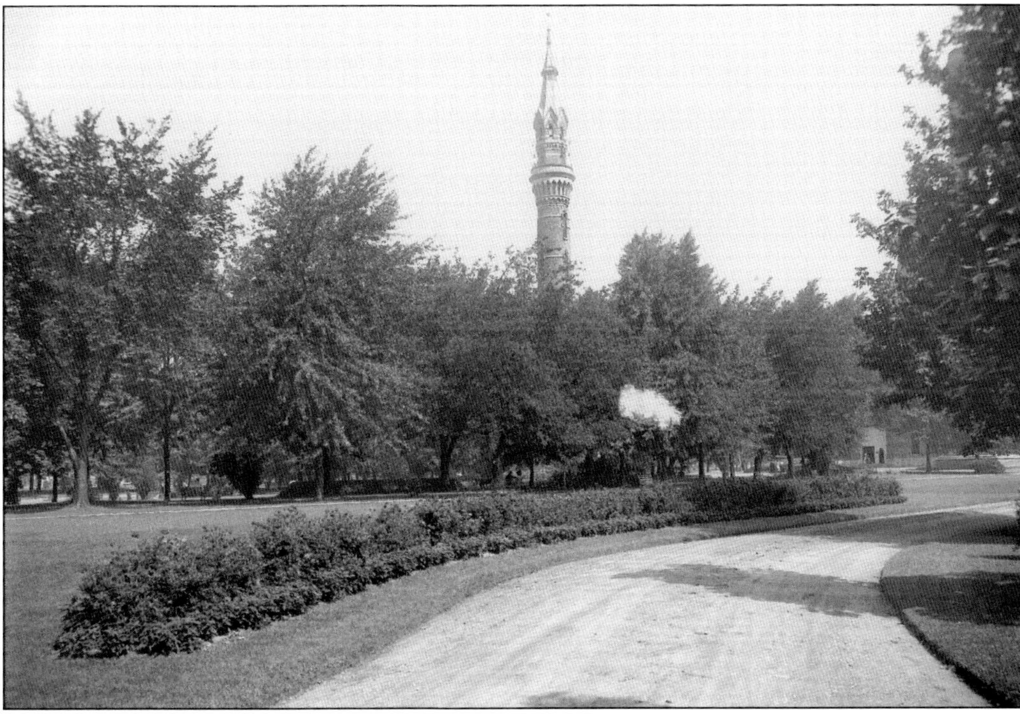

The driveway paralleled the old coaling canal along its entire length at Water Works Park. The image above is near the top of the canal about 1920. The image below, from 1904, is closer to the foot of the park, at the Detroit River. The trees lining the canal consist of a variety of species, including chestnuts and apples, among others. The ancient French pear trees known as the Twelve Apostles are located in the vicinity of the original pumping station and standpipe in the distance. The Apostles date from the French occupation of Detroit in the early 18th century and had died out by the mid-1940s. (Above, DPC/LOC; below, author.)

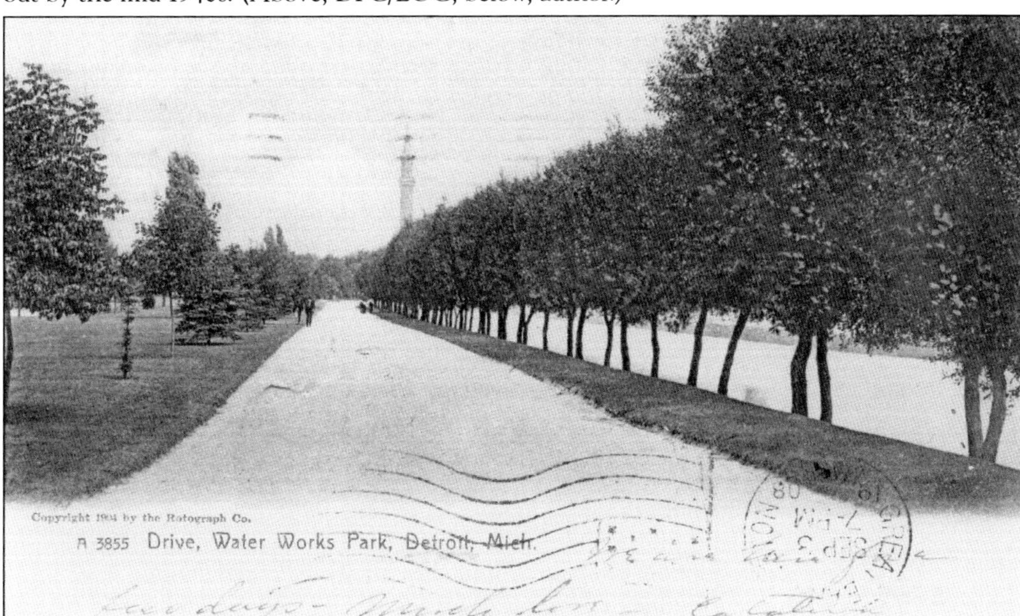

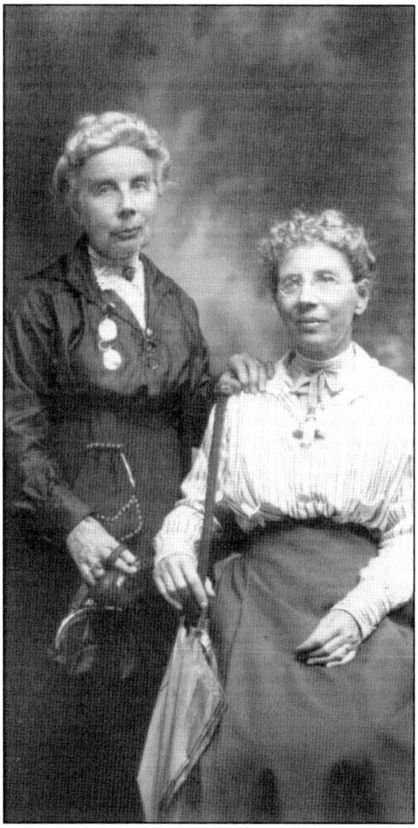

Sylvia Allen (left) operated a private kindergarten for children of prominent Detroiters with her sister, Julia. Sylvia, who died in December 1929, left her estate to the City of Detroit "to establish and maintain a garden and playground for children in a congested neighborhood in the City of Detroit." In 1940, the city designated the former Ladder Company 14, by then vacant, the Sylvia Allen Recreation Center. The facility was a hobby center specializing in the teaching of crafts. It operated until 1968. The Allen sisters' grandfather Gen. Charles Larned, a veteran of the War of 1812 and attorney general of the Michigan Territory, is Larned Avenue's namesake. (Burton Collection.)

The Detroit Fire Department's Ladder Company No. 14 originally occupied this building on July 15, 1912. Increasingly larger firefighting equipment, however, made the structure obsolete for firefighting purposes by 1918. The building sat vacant until 1940, when the lease was transferred to the Detroit Recreation Department, which used it to house the Sylvia Allen Recreation Center. Following the closing of the center in 1968, the lease reverted to the fire department, which used it for storage. After the building was returned to the water department in 1987, arrangements were made for it to be used by Detroit's historic Pewabic Pottery. (DWSD.)

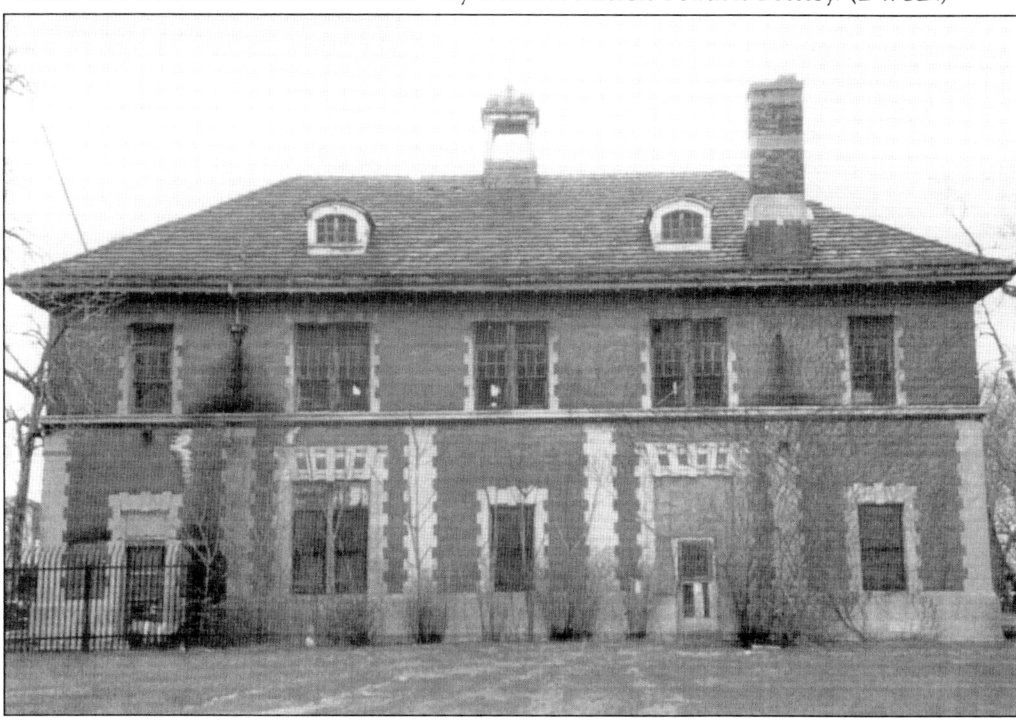

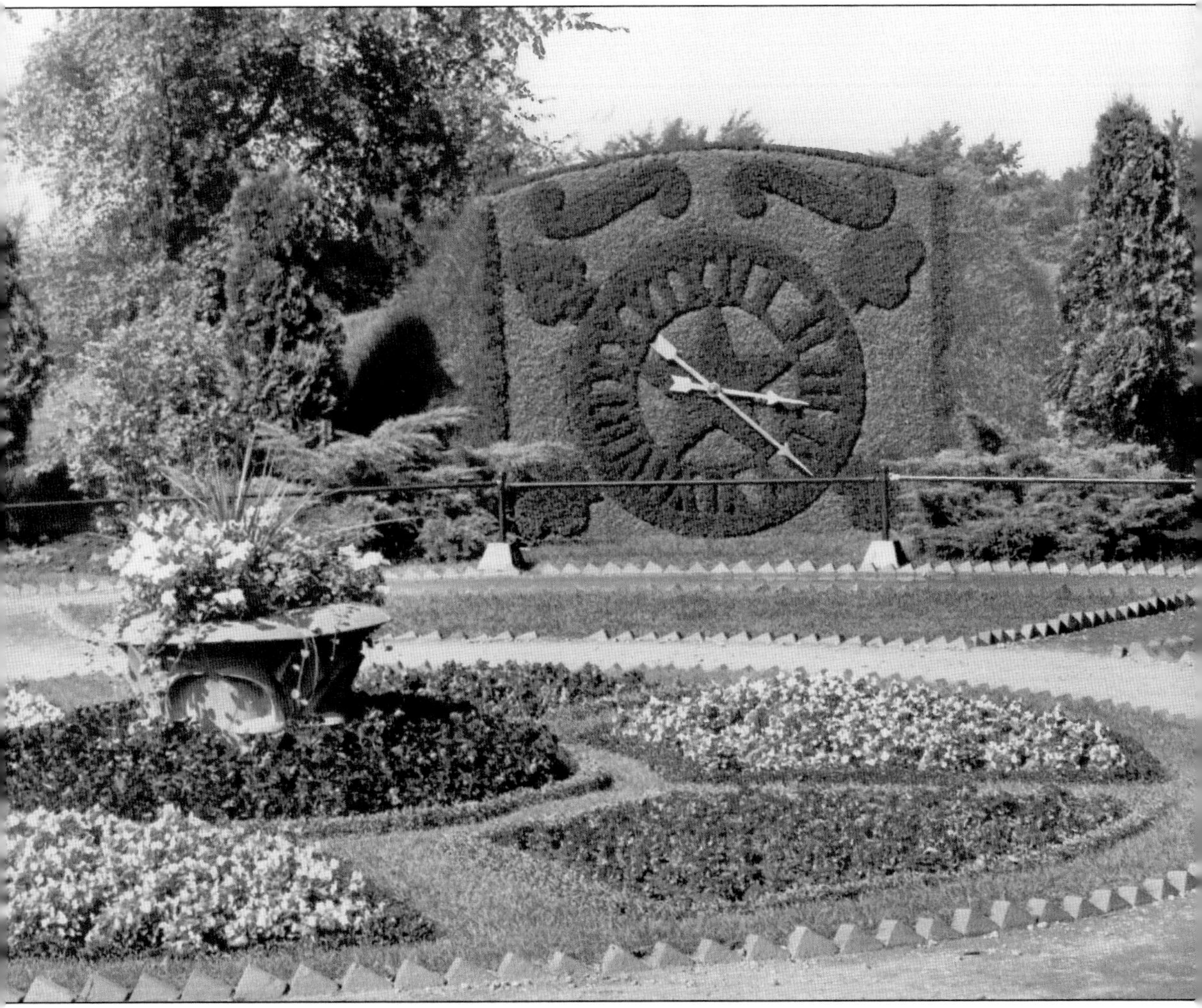

The floral clock, created by Elbridge A. Scribner, was hydraulically powered by water surging beneath the park. The clock became a signature attraction after its debut in 1893. This image shows the clock after it was moved to Henry Ford's Greenfield Village in 1934. (DWSD.)

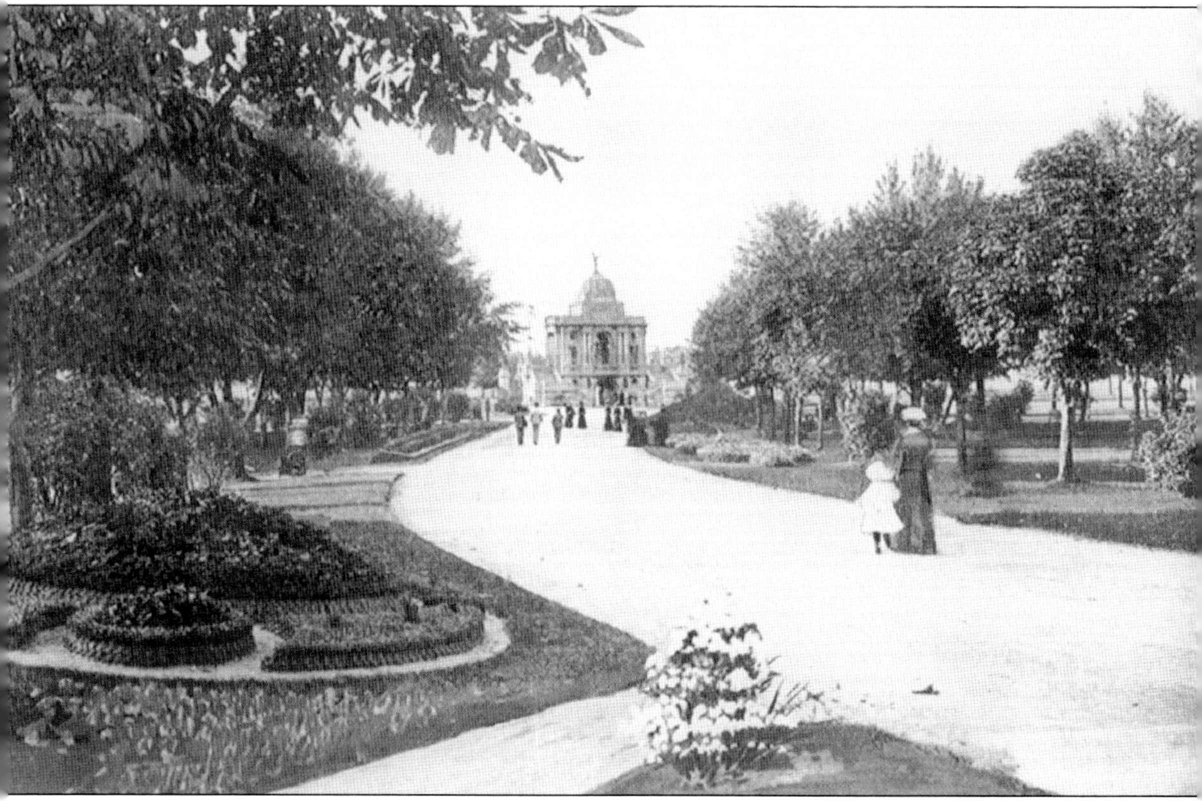

Seeing visitors file past floral displays as they walk out through the Hurlbut Memorial Gate is an appropriate metaphor for Water Works Park. Despite hopeful talk about reopening the park to the public, the likelihood of the return of free-roaming visitors to the grounds in a post-9/11 world is likely to remain unrealized speculation. (Author.)

Three

Defender of Public Health

With its peaceful, serene grounds, Water Works Park was one of Detroit's premier public parks for many years. Because of its enormous popularity and many images of visitors engaged in leisurely activities, it is easy to lose sight of the park's primary function as a supplier of drinking water. In a way, the public water utility seems to have been hiding in plain sight.

Thanks to the Great Lakes, the world's largest concentration of fresh water, Detroit has always been blessed with an abundance of unusually pure water. However, the increasing density of population in southeast Michigan in the 19th century, combined with the common practice of discharging raw sewage and industrial wastes into the nearest body of water, made consumption of drinking water drawn from the Detroit River increasingly deadly.

Drinking tap water in Detroit and other large American cities became riskier in the late 1800s. Deadly *Escherichia coli*, or E. coli—a rod-shaped bacterium prevalent in fecal matter that originates in the lower intestines of warm-blooded organisms—was the culprit. Death from typhoid fever in the United States occurred at a rate of 36 per 100,000 people in 1899. From November 1, 1899 to October 31, 1900, there were 65 deaths from typhoid in Detroit, or 22.7 per 100,000 residents. It would get progressively worse.

In 1912, Water Works Park became ground zero in Detroit's fight against typhoid fever and other deadly waterborne illnesses. The US Public Health Service, a forerunner of the Environmental Protection Agency, mandated that Detroit's drinking water be disinfected to kill deadly pathogens, especially E. coli, the first waterborne contaminate to be regulated.

Sodium hypochlorite, a bleach-like solution, was added to the drinking water distributed from Water Works Park beginning in 1913. The results, however, were mixed. It was after the introduction of liquid chlorine, a more powerful disinfectant, in 1916, that deaths from waterborne disease dropped dramatically. Because of similar reductions in other cities, liquid chlorine has remained the gold standard of water disinfection ever since.

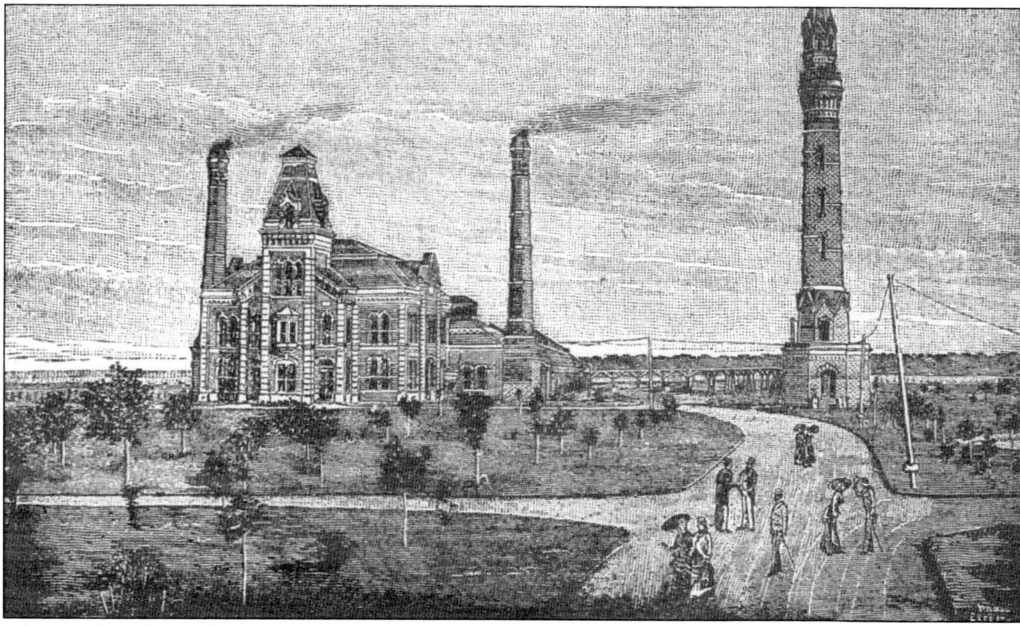

This undated woodcut depicts the original pumping station at Water Works Park in its initial configuration, with a mansard roof. Contrary to the information in the woodcut, station operations actually began in 1879, not 1873. The building was modified and expanded in 1886, and again in 1900, to accommodate more steam engines. The mansard roof was replaced with a more conventional style roof during renovations performed from 1900 to 1903. (DWSD.)

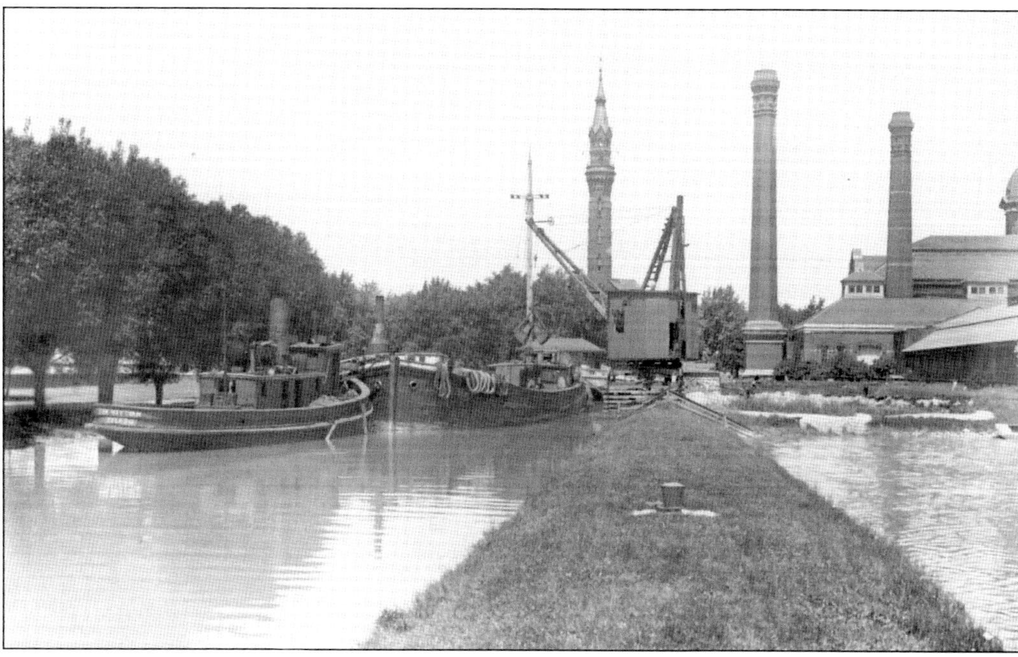

In 1905, the canal was dredged to accommodate larger vessels. The water to the right is a portion of a settling basin used prior to the construction of the first water treatment plant in 1923. Slow movement of water in settling basins allowed dirt and large particulate matter to settle out to the bottom. (DWSD.)

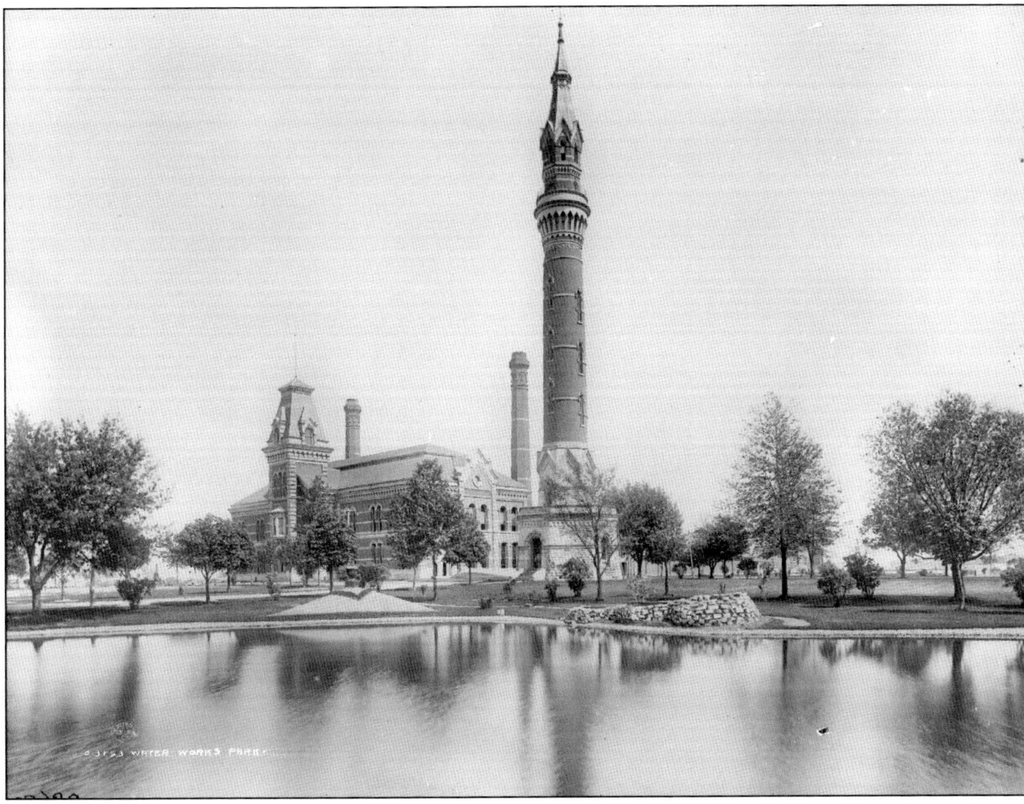

The above image shows the pumping station following its 1886 expansion. The steam-pumping engine installed that year was designed by John Edwards, chief engineer of the waterworks, and built by Samuel F. Hodge & Company, a Detroit-based manufacturer of marine engines founded by former Detroit water commissioner Samuel F. Hodge. For several decades, it was the world's largest steam engine. That same year, the Detroit Board of Water Commissioners purchased the Old Fireman's Hall at Randolph Street and East Jefferson Avenue in downtown Detroit. It served as the board's headquarters until 1928, when it was replaced by the current Water Board Building. (Above, DPL/LOC; below, DWSD.)

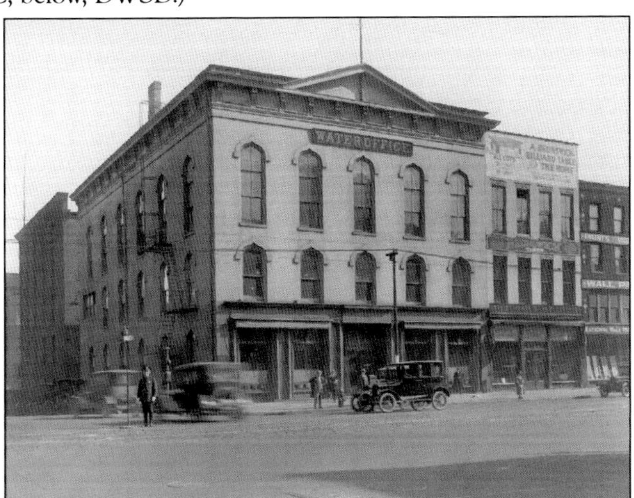

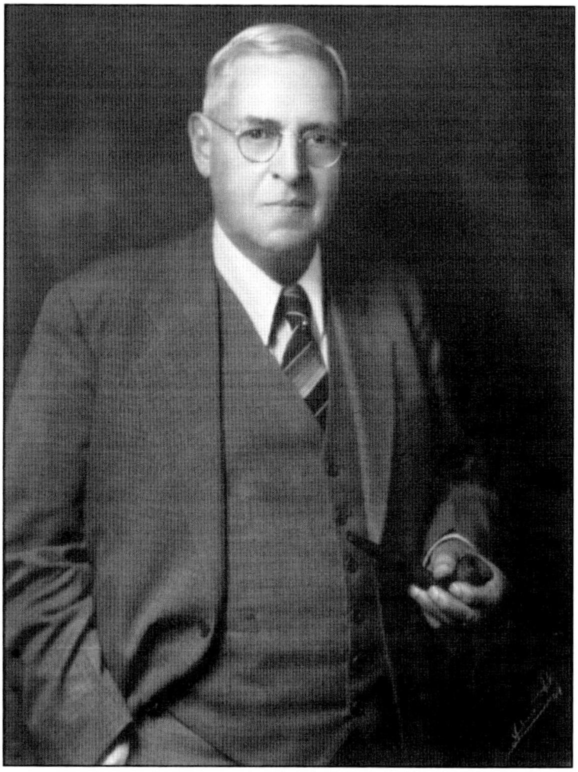

As the Detroit Board of Water Commissioners' engineer-in-charge, Clarence W. Hubbell designed the Detroit River water intake that fed Water Works Park, shown below, in 1905. As the city engineer in the 1910s, Hubbell designed and oversaw construction of a new state-of-the-art sewer system for Detroit. He authored important studies of pollution in the Detroit River in 1916 and 1925 that were instrumental in the establishment of the sewage treatment plant in Detroit in 1940, still the world's largest sewage treatment plant on a single site. In 1932, he designed the water intake at the head of Belle Isle that would be used to feed Detroit's new Springwells Water Treatment Plant in 1935. Hubbell also designed the Northeast Water Treatment Plant in 1956. (Left, Hubbell, Roth & Clark, Inc.; below, DWSD.)

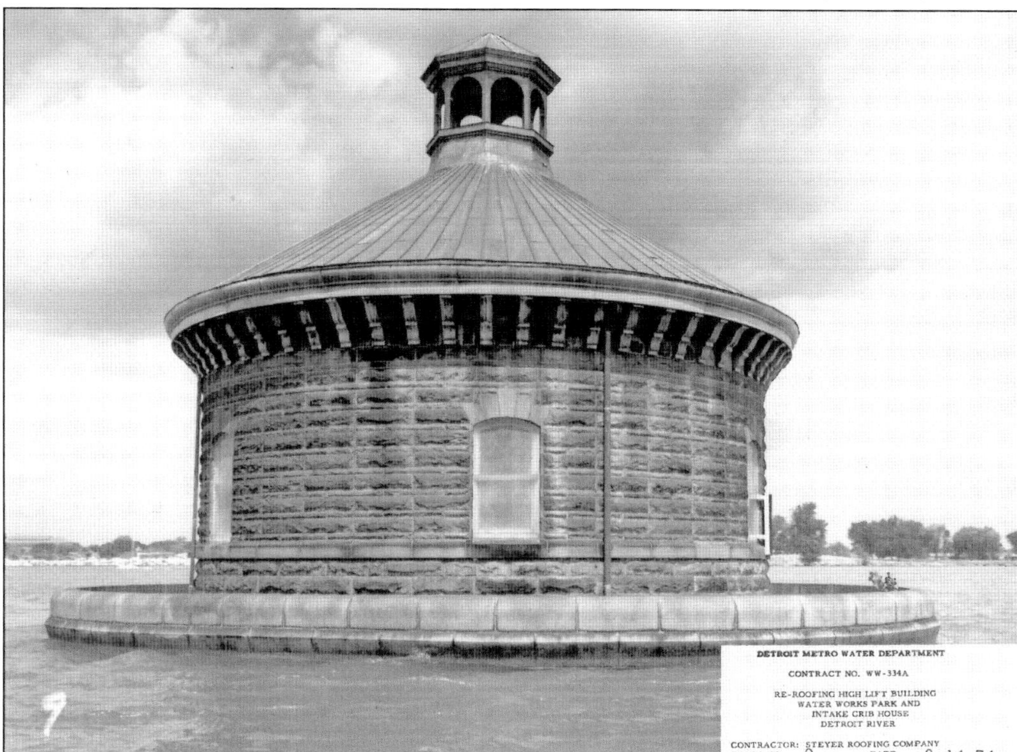

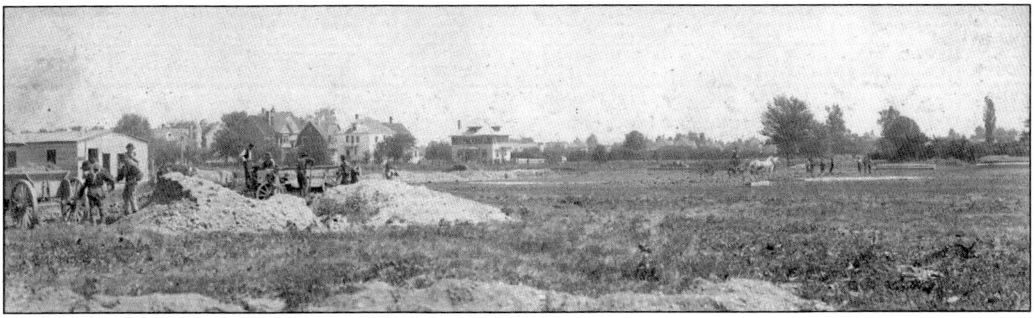

The site of the new pumping station (currently known as the High Lift Building) at Water Works Park is seen above at the start of construction on September 2, 1909. The below photograph was taken only a few weeks later, when construction had begun in earnest. Although Detroit was commonly known as Motor City by 1909, horses remained a familiar sight on the city's streets, most of which remained unpaved. Heavy construction was still greatly reliant on the broad shoulders and strong backs of men with shovels, and horses. (Both, DWSD.)

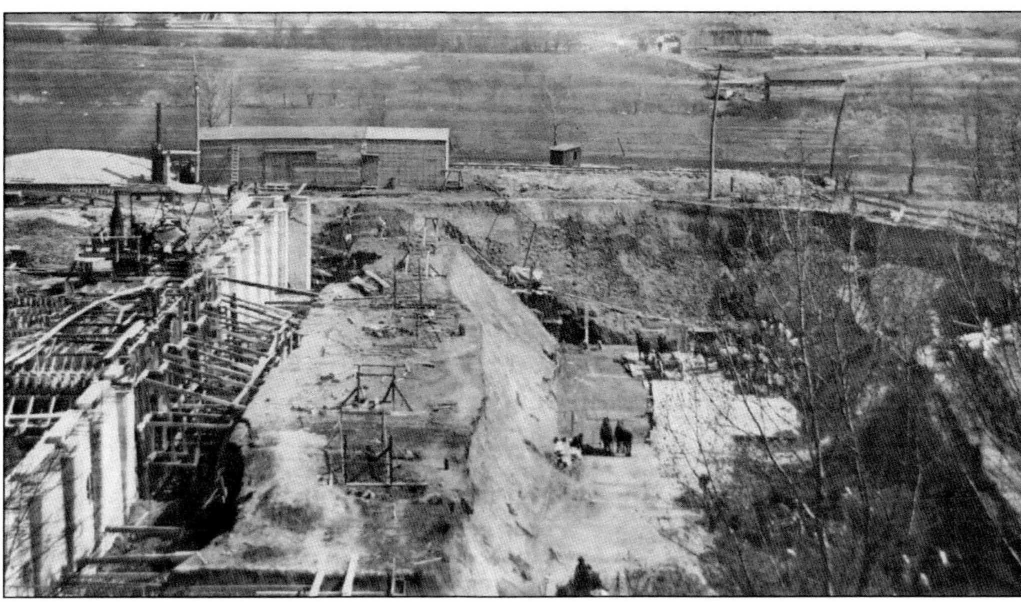

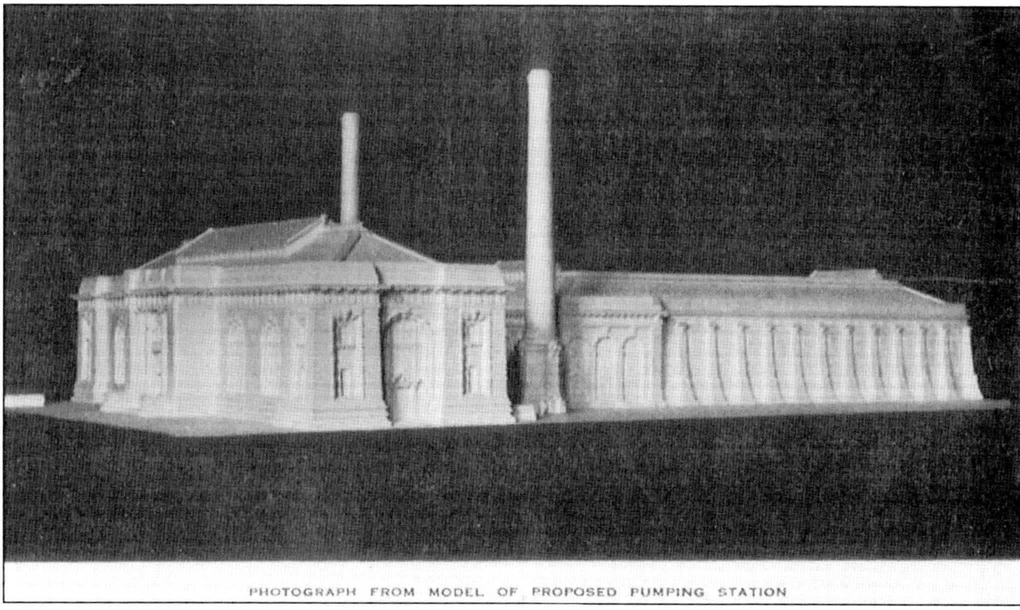

The need for more water in Detroit was an ongoing problem in the years before and after 1900. Demand was generated by the city's rapid development into an industrial powerhouse, led by the exploding growth of automobile manufacturing. Despite a major expansion from 1900 to 1903 that doubled the size and output capacity of the original pumping station, the Detroit Water Department struggled to supply increasing demand. A model of the new pumping station proposed for Water Works Park in about 1907 is shown above, while the image below shows construction underway on the station around 1909. (Both, DWSD.)

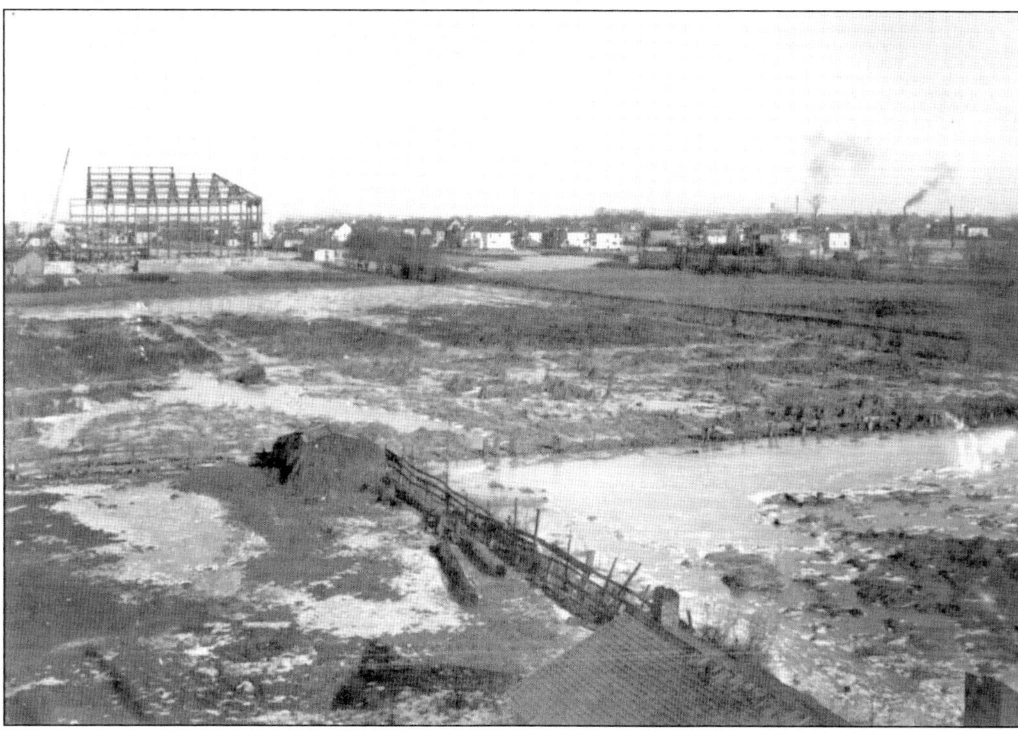

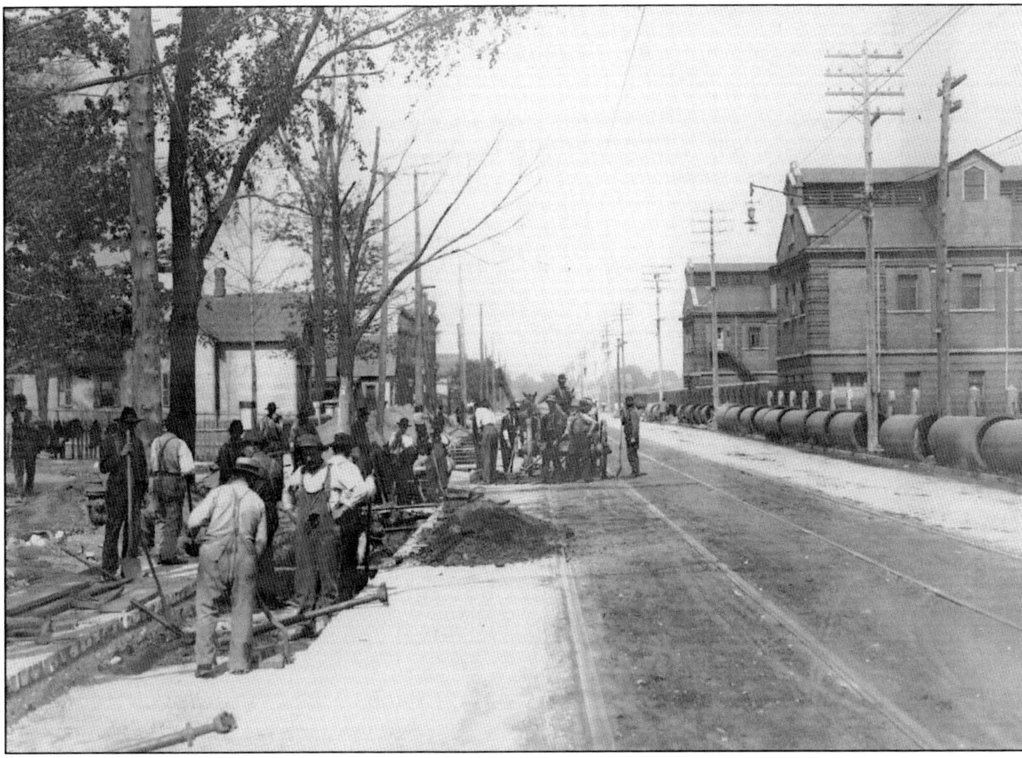

Workers laying distribution water mains in July 1908 provide a vivid portrait of the extremely labor-intensive effort necessary to bring water into homes and businesses prior to the use of more specialized heavy equipment. (DWSD.)

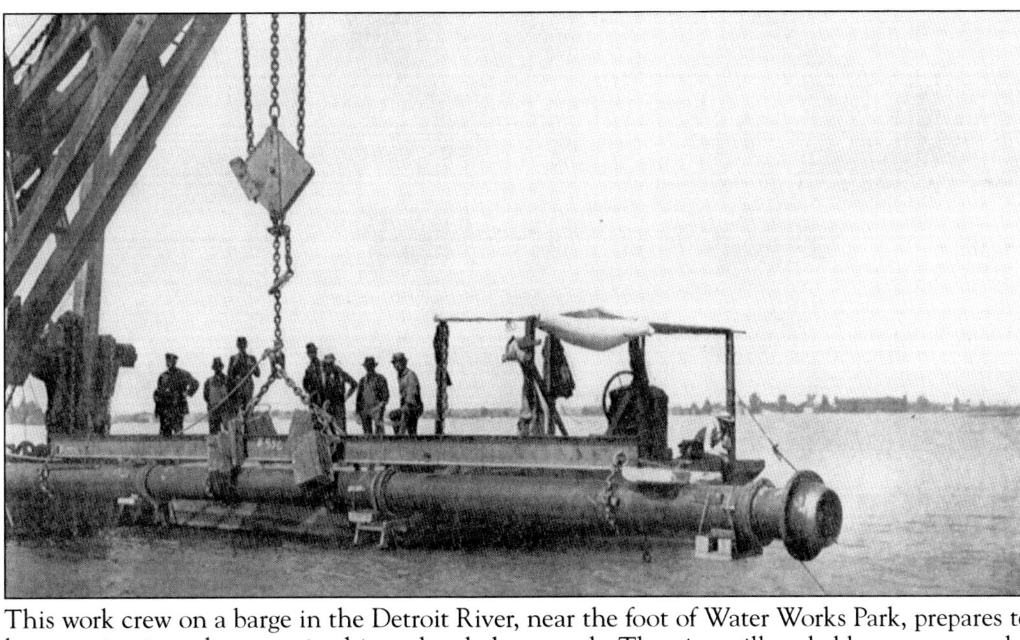

This work crew on a barge in the Detroit River, near the foot of Water Works Park, prepares to lower a pipe into the water in this undated photograph. The pipe will probably connect to the intake, supplying raw river water to the park. (DWSD.)

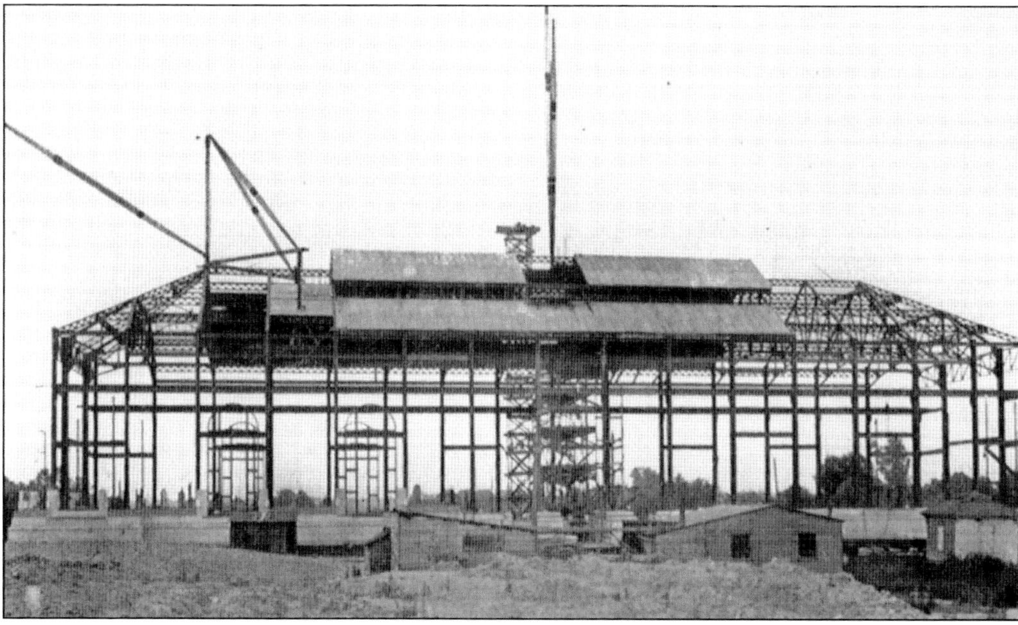

Pumping Station No. 2 is seen here both during construction (above) on June 20, 1911, and as it appeared from Jefferson Avenue in 1936 (below). The station began pumping water into Detroit's water distribution mains in 1914. Despite the additional capacity it generated, the city's water department struggled to keep pace with the increasing demands of Detroit industry led by Henry Ford, his contemporaries at General Motors, the Dodge brothers, and countless others. This facility would witness and survive the transition from coal-fired, steam-generated power to electricity that occurred from 1959 to 1962. Detroit's iconic Pumping Station No. 2, currently known as the High Lift Building, is a hodgepodge of different architectural styles including Egyptian, Roman, Beaux Arts, and Art Deco. (Both, DWSD.)

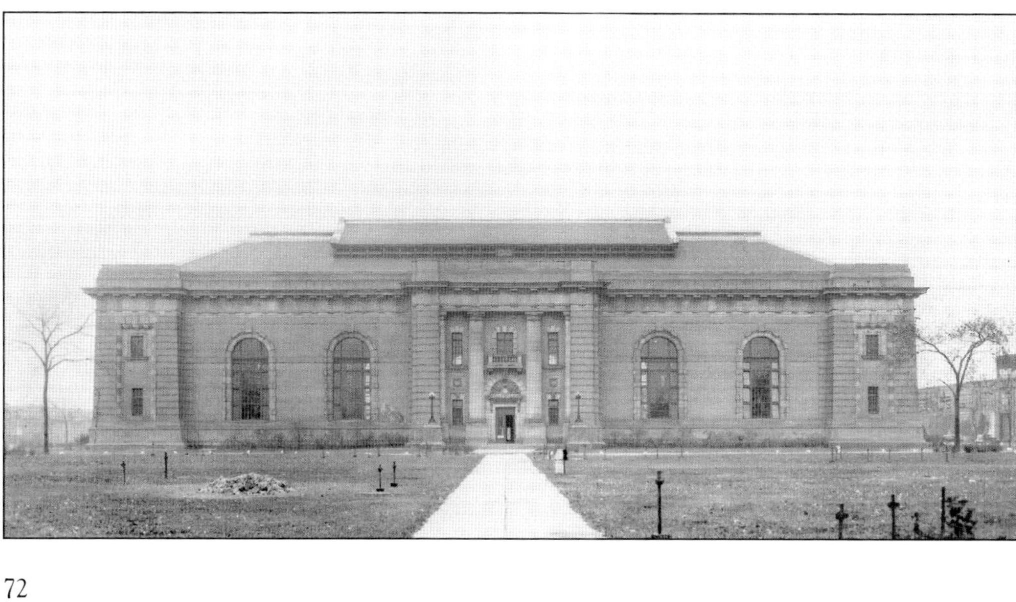

The disinfection of drinking water was a personal matter to Benjamin Franklin Guiney (right). Guiney, who briefly played major league baseball as a catcher for the Detroit Wolverines of the old National League, lost his son, Karl Guiney (below), during one of the periodic outbreaks of typhoid fever in Detroit. The elder Guiney was a member of the Detroit Board of Water Commissioners briefly in 1900 and then from 1914 to 1919, a period when Detroit was fully engaged in a struggle to overcome high mortality rates from typhoid fever and other water-borne diseases. Detroit began disinfecting its public drinking water with sodium hypochloride in 1913 at the direction of the US Public Health Service. Liquid chlorine, a more powerful disinfectant, has been used since 1916. (Both, the Guinney family.)

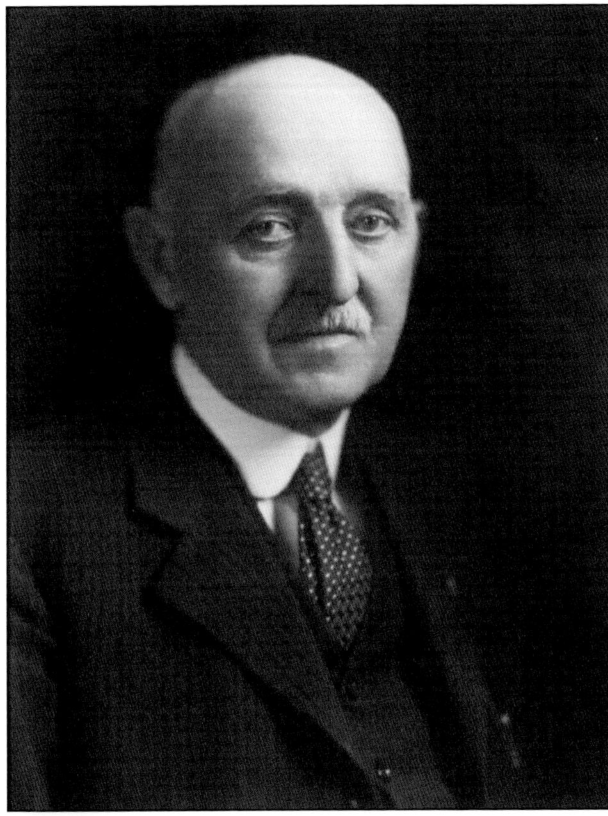

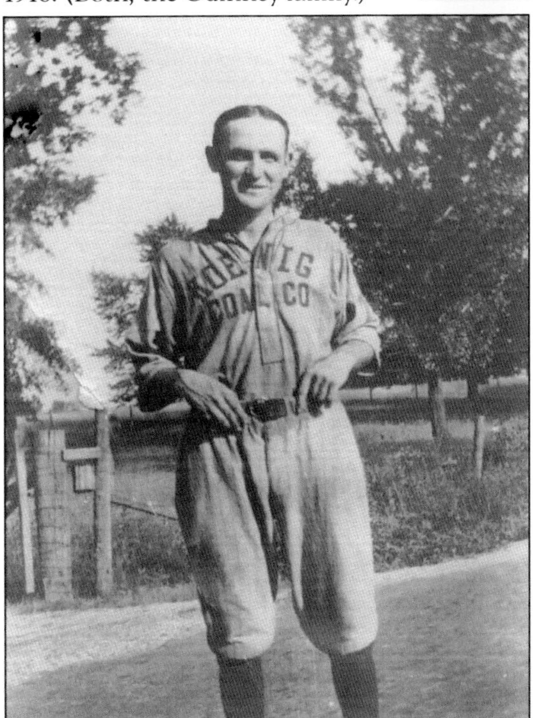

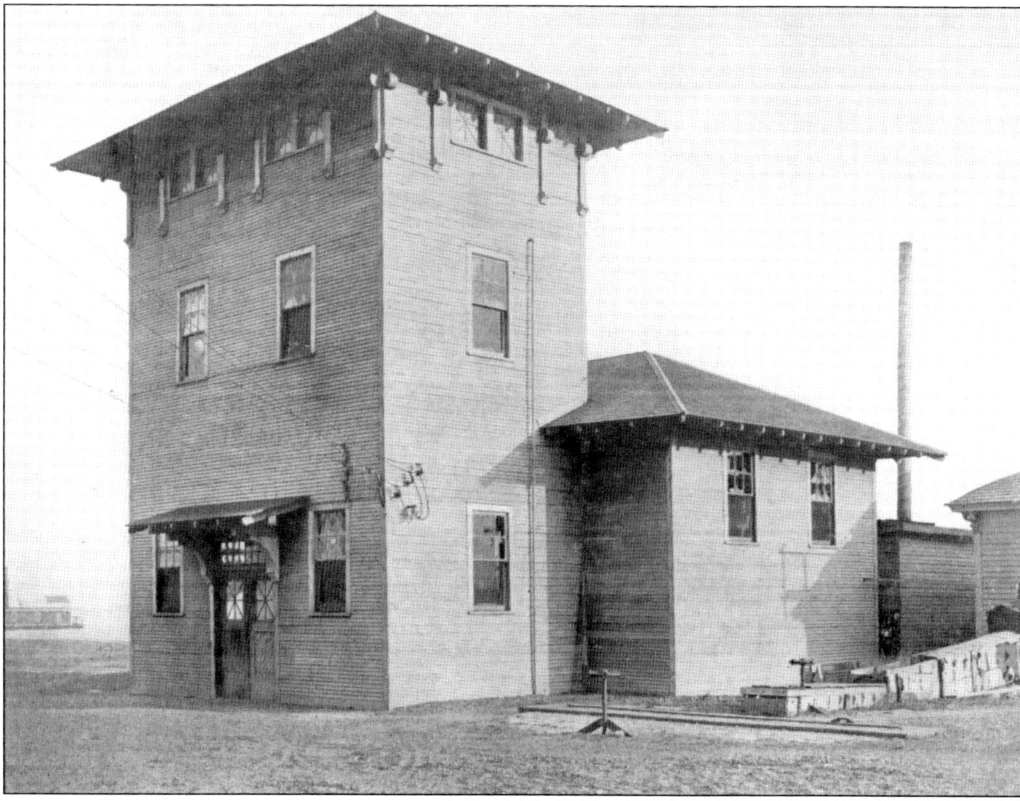

Before approving costly construction of a new water treatment plant, the Detroit Board of Water Commissioners wanted to assess Detroiters' reaction to colorless filtered water. People had been accustomed to drinking a less-than-transparent beverage, due primarily to the presence of sediment. An experimental filtration plant was built in Water Works Park (above), and commissioners intended to give the public access to the inside of the plant (below) to see the filtration process in action. When the United States entered World War I, however, Water Works Park was closed to the public for reasons of national security. To compensate, the plant's filtered-water output was piped to a tent next to the Hurlbut Memorial Gate, where the public was invited to compare the treated and filtered water to their normal, unfiltered drinking water. (Both, DWSD.)

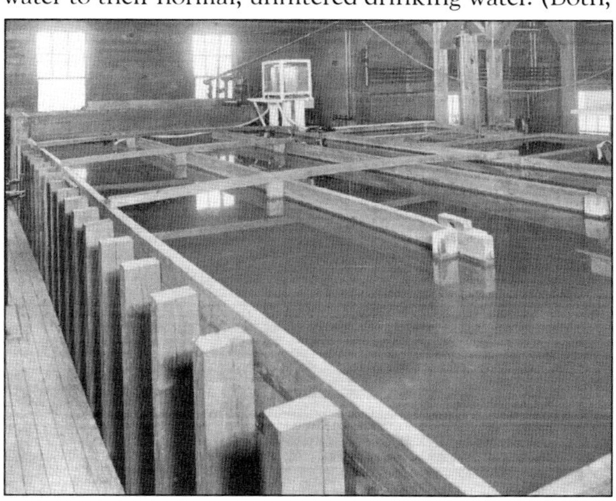

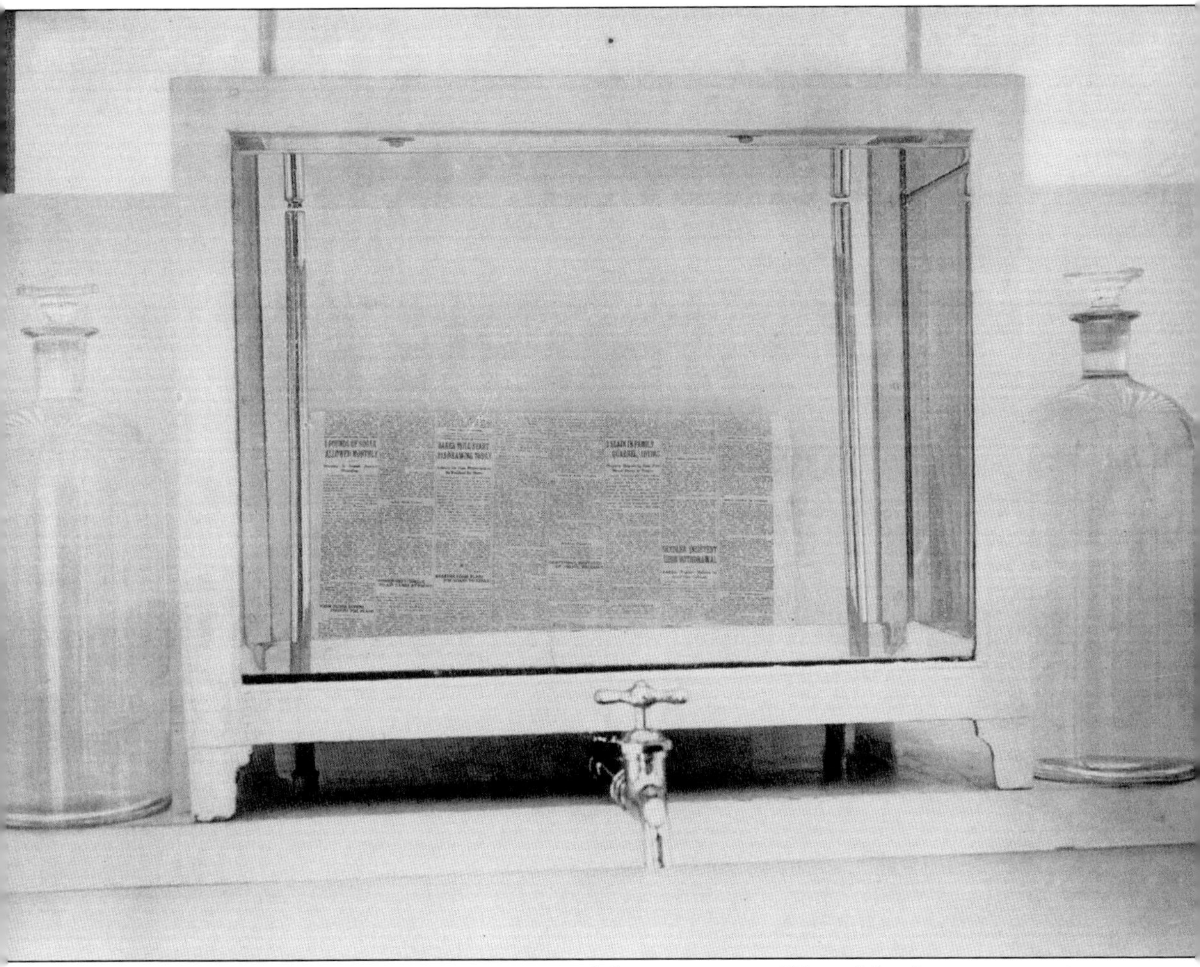

The Detroit Board of Water Commissioners touted the superiority of filtered drinking water, saying it was "so clear you can read a newspaper through it." Previously, the unfiltered drinking water was discolored by sediment, and sometimes contained other objects, including bugs. (DWSD.)

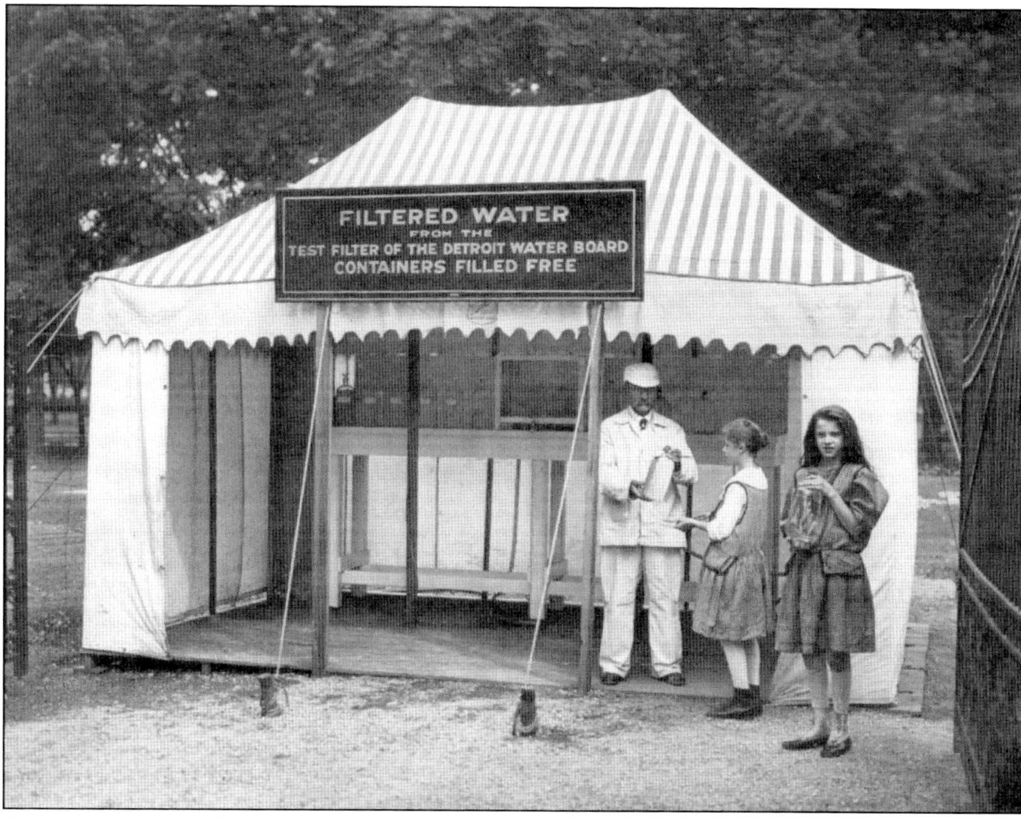

A white-suited employee of the Detroit Board of Water Commissioners offers free samples of filtered drinking water to a pair of young girls. Detroiters wholeheartedly embraced the newfangled transparent water. As a result, the Detroit Board of Water Commissioners approved construction of the permanent treatment plant at Water Works Park. (DWSD.)

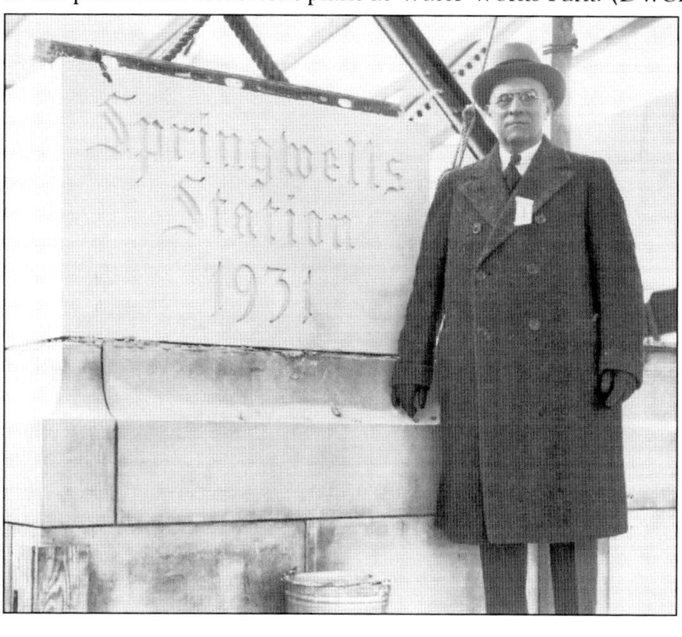

George Harrison Fenkell poses next to the cornerstone of the Springwells Station, Detroit's second water treatment plant, in February 1931. From 1918 to 1936, Fenkell served as director of the Detroit Water Department, during which time Detroit's role as a service provider greatly expanded. It was during his administration that the city's first two water treatment plants were established, and work began on the Detroit Wastewater Treatment Plant that is still the world's largest single-site facility. (DWSD.)

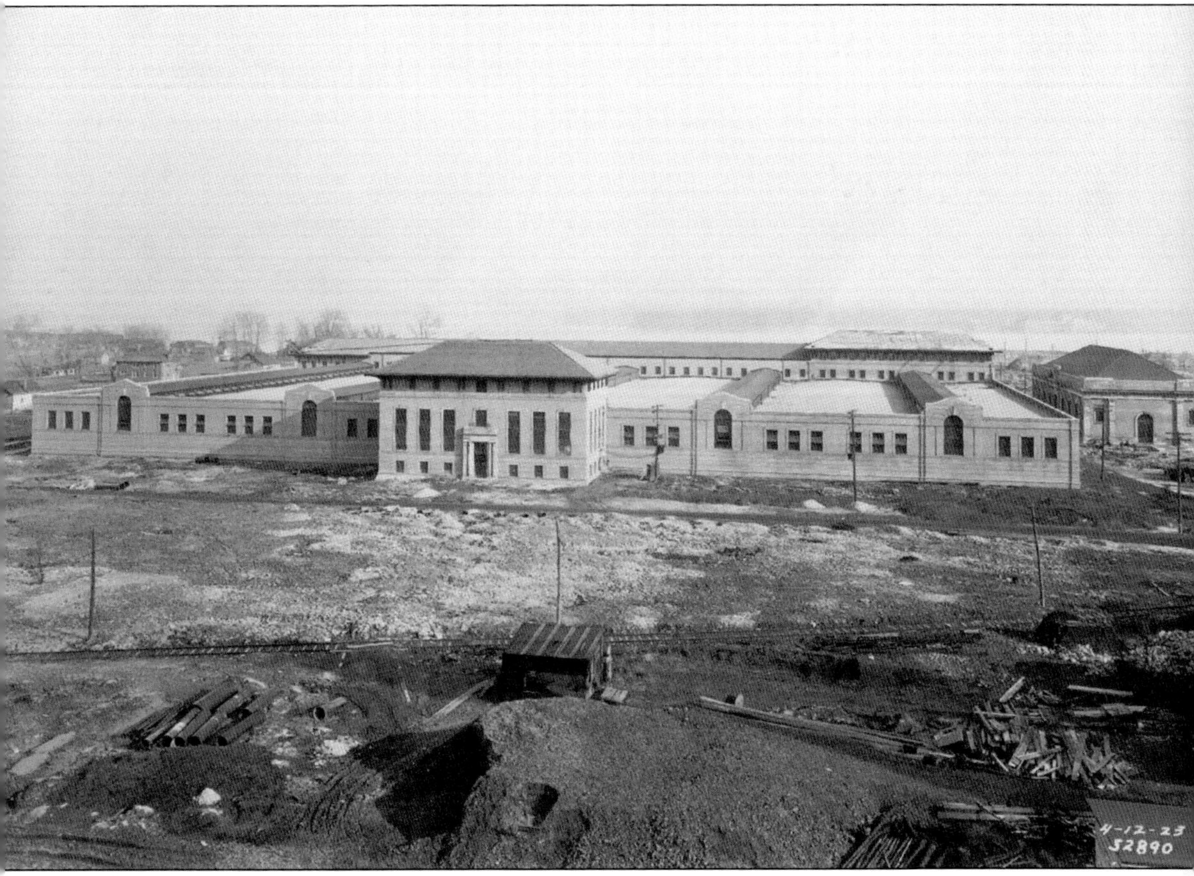

The water treatment plant at Water Works Park was nearly completed in April 12, 1923. This was Detroit's first water treatment plant and faithfully served the city's water drinkers for 80 years. The plant was eventually replaced by a new water treatment plant, which entered service in 2003. (DWSD.)

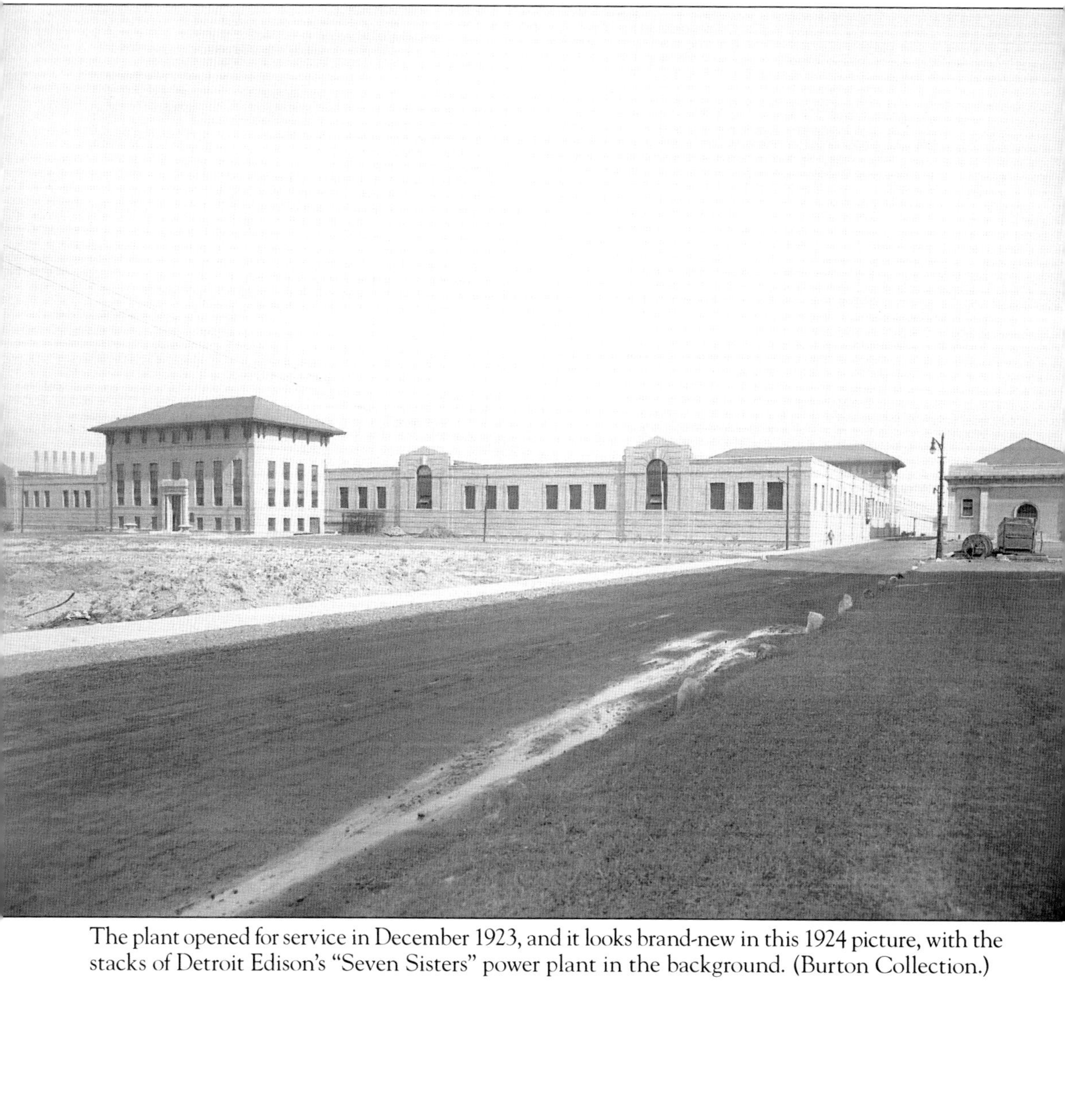

The plant opened for service in December 1923, and it looks brand-new in this 1924 picture, with the stacks of Detroit Edison's "Seven Sisters" power plant in the background. (Burton Collection.)

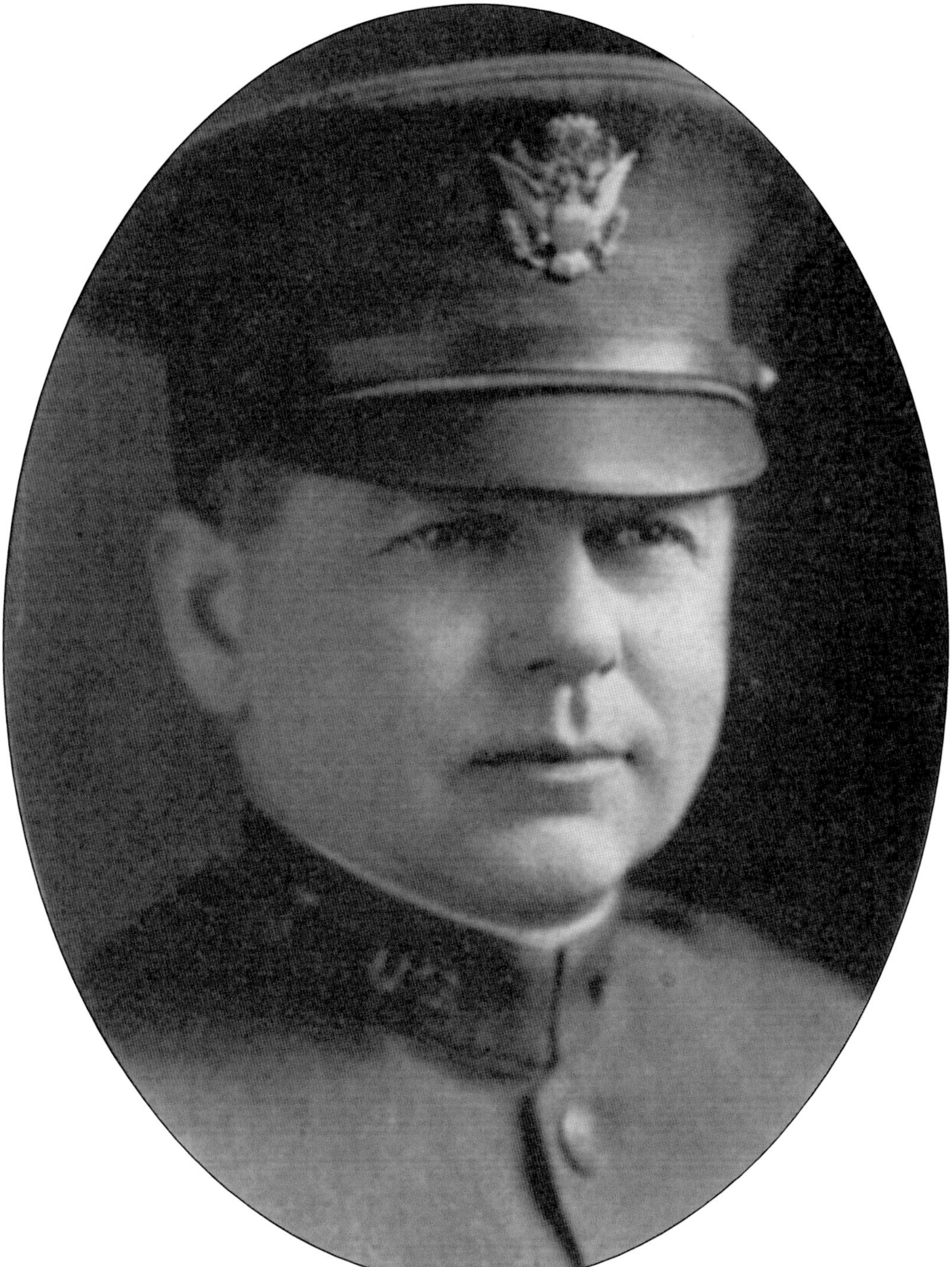

Theodore A. Leisen, P.E., designed and supervised construction of the water treatment plant. Leisen (shown here in his World War I–era Army uniform) served as a civil engineer for the Detroit Water Department until 1924. He then became general manager for the Municipal Utilities District in Omaha, Nebraska. In 1993, Leisen was inducted into the Water Industry Hall of Fame. (DWSD.)

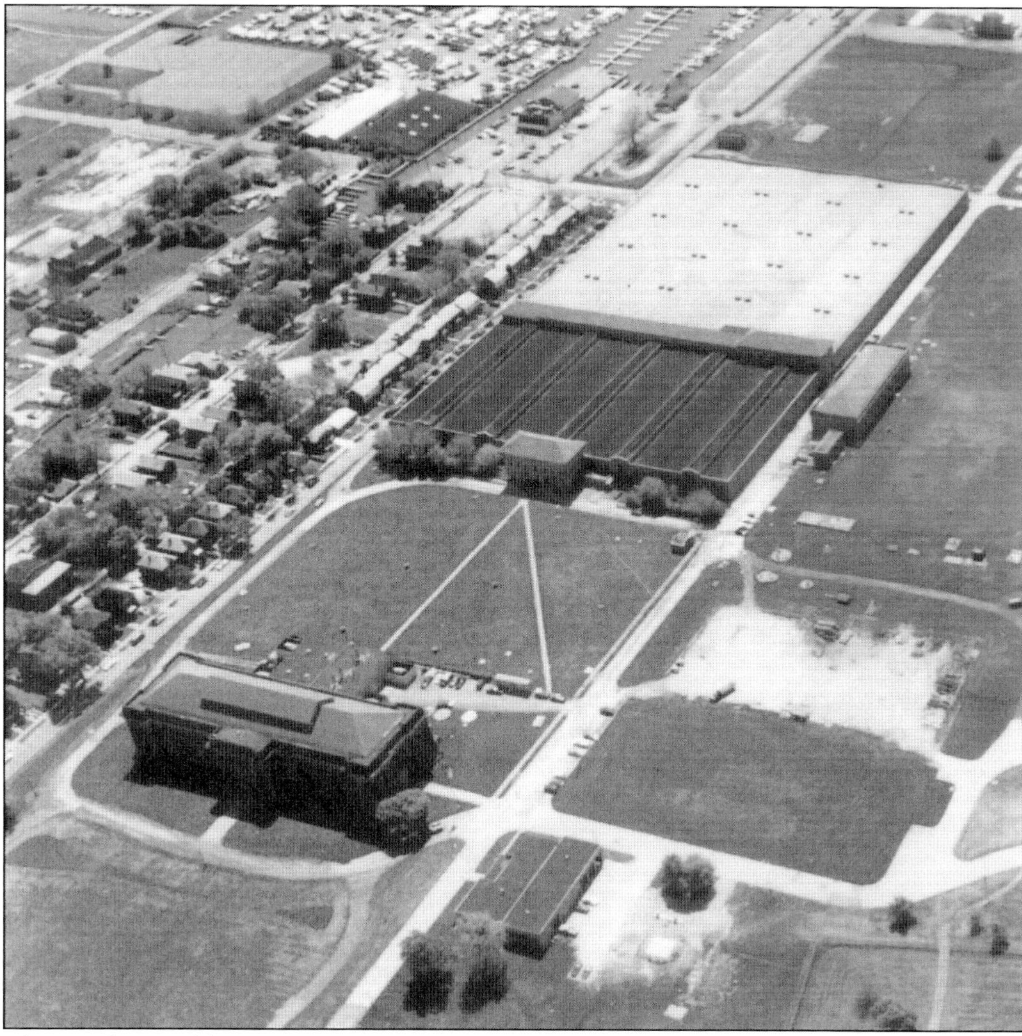

This aerial view of the water treatment plant from the 1960s or 1970s shows the High Lift Building in front of the plant near the bottom. The original pumping station sat on the vacant area to the right of the High Lift Building. (DWSD.)

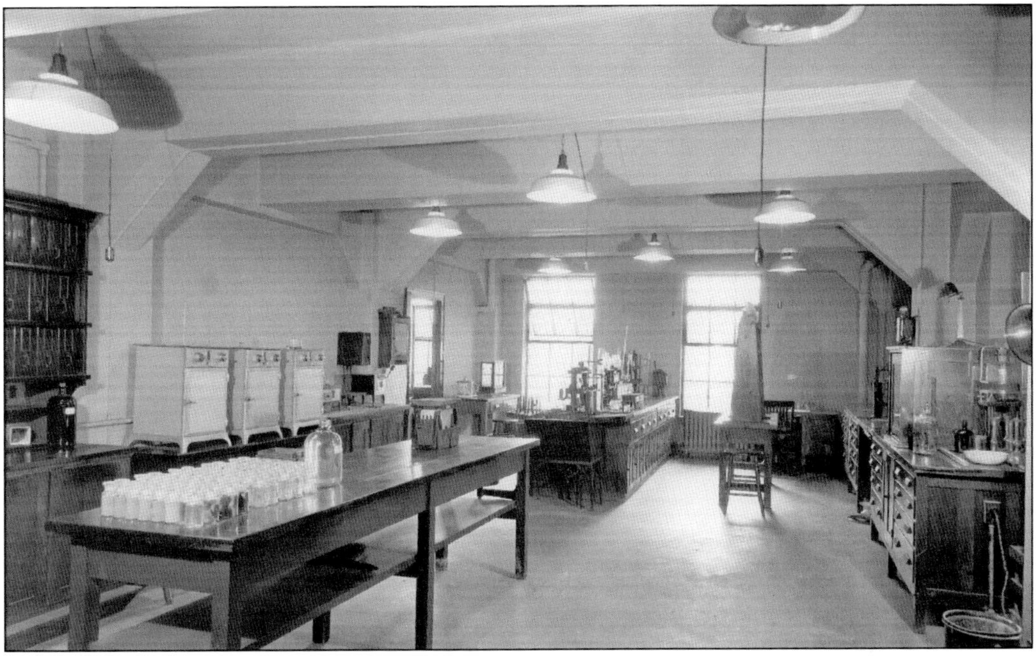

An unidentified chemist tests for the presence of pathogens in water samples (below) in the chemistry lab at Water Works Park (above). Disinfection of drinking water has been credited for the 50-percent increase in life expectancy experienced in the United States in the 20th century, as well as one of the most significant advances in public health of the last 100 years. Disinfection, mostly by chlorination, was not, however, an immediate success. It was a largely hit-or-miss application until formulas were adjusted to find the right balance. In Detroit, the introduction of 24-hour monitoring by "degreed chemists" to ensure consistency of chlorination was largely responsible for the virtual eradication of illness and death from typhoid fever and other waterborne disease by the mid-1920s. (Both, DWSD.)

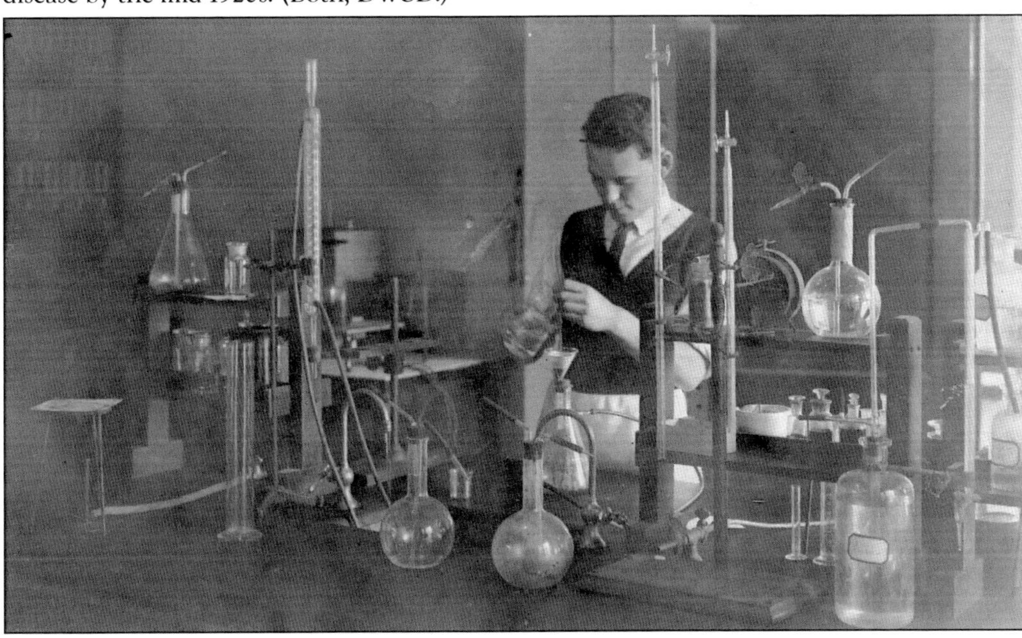

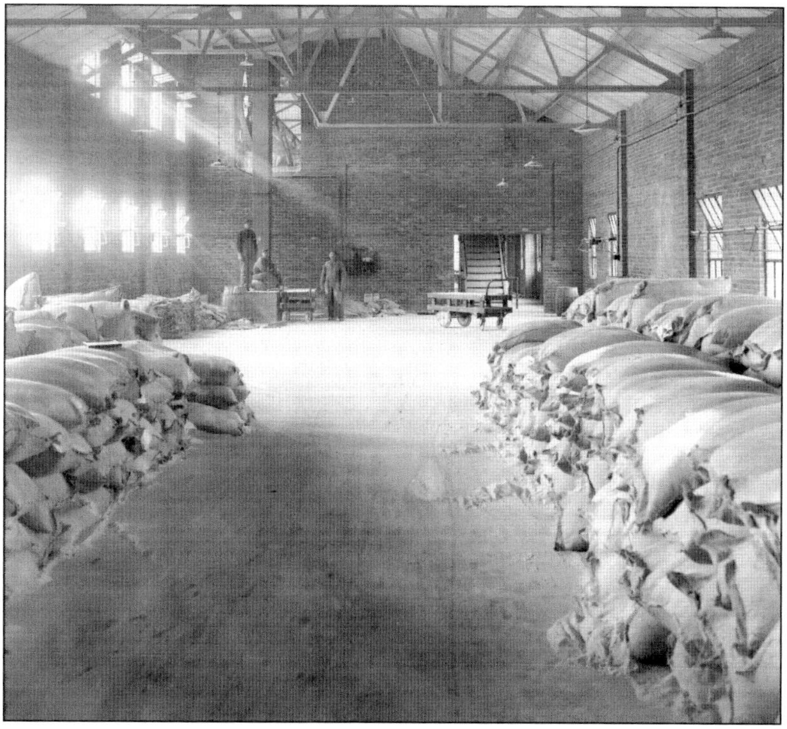

This storage room kept alum, an aluminum sulfate mixture, in the water treatment plant in the 1920s. Commonly used as a coagulating agent in water treatment, alum attracts dirt and other particulate matter suspended in raw water, and forms large flakes known as floc. Being heavier than the surrounding water, the floc settles to the bottom of settling basins before being removed and disposed. (DWSD.)

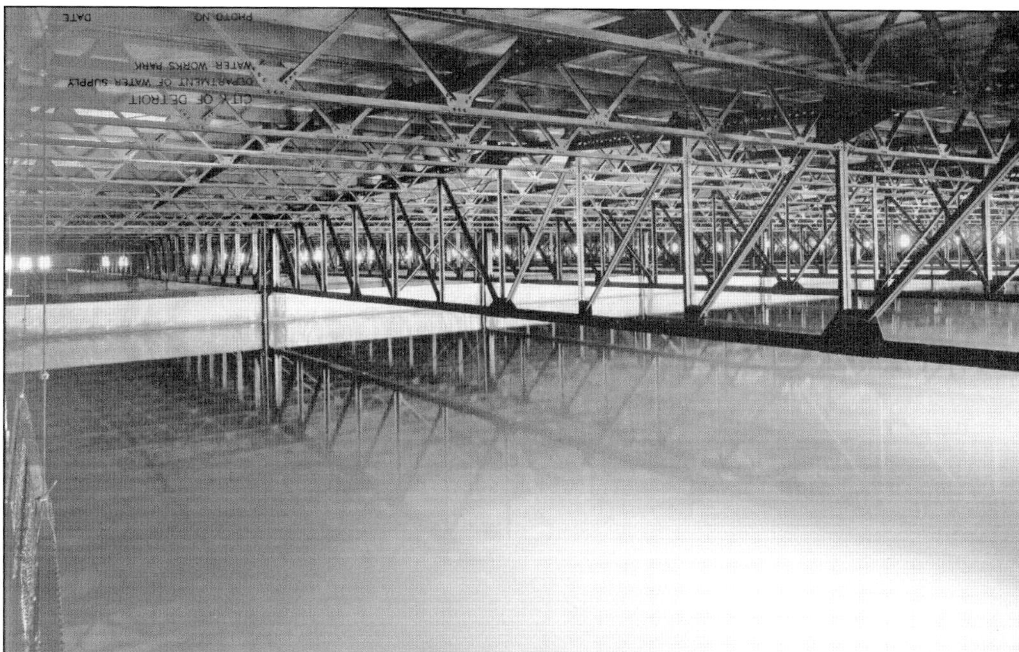

This undated image shows the coagulation and sedimentation basin at Water Works Park, which is actually shown right side up. Water entered this basin after the addition of alum. Dirt and other impurities were attracted to the floc, creating heavier particles that settled to the bottom of the basin. The cleaner water remaining was then drawn—or decanted—from the top and directed to the next stage of the treatment process. (DWSD.)

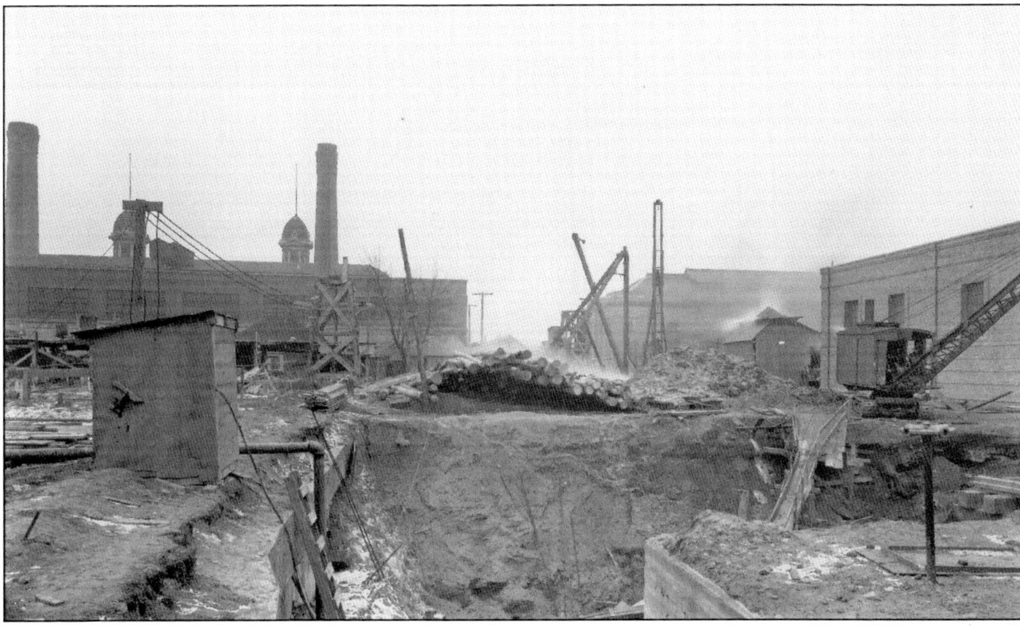

Despite being the world's largest upon entering service in December 1923, the water treatment plant struggled to keep pace with the ravenous demands of Detroit's automakers and myriad other industrial concerns. Within six months, the Detroit Board of Water Commissioners and the city council approved construction of a second water treatment plant in nearby Springwells Township (now Dearborn). In the meantime, board members had to do what they could to squeeze more water from the existing plant, already operating at maximum capacity. The image above shows construction of a new underground water reservoir at the park in December 1926. The image below is of a water board work crew on the job in October 1928. (Both, DWSD.)

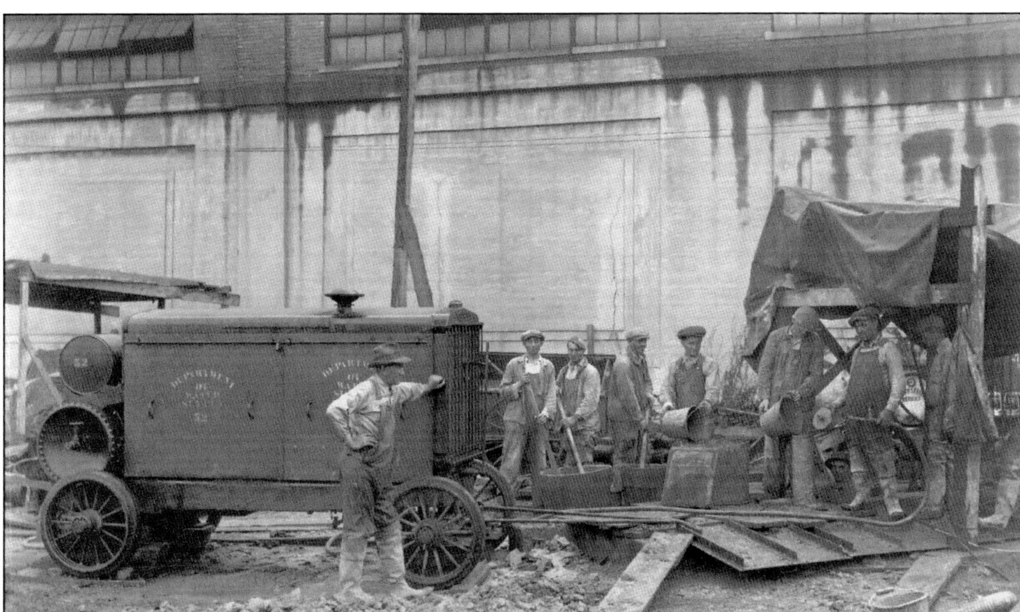

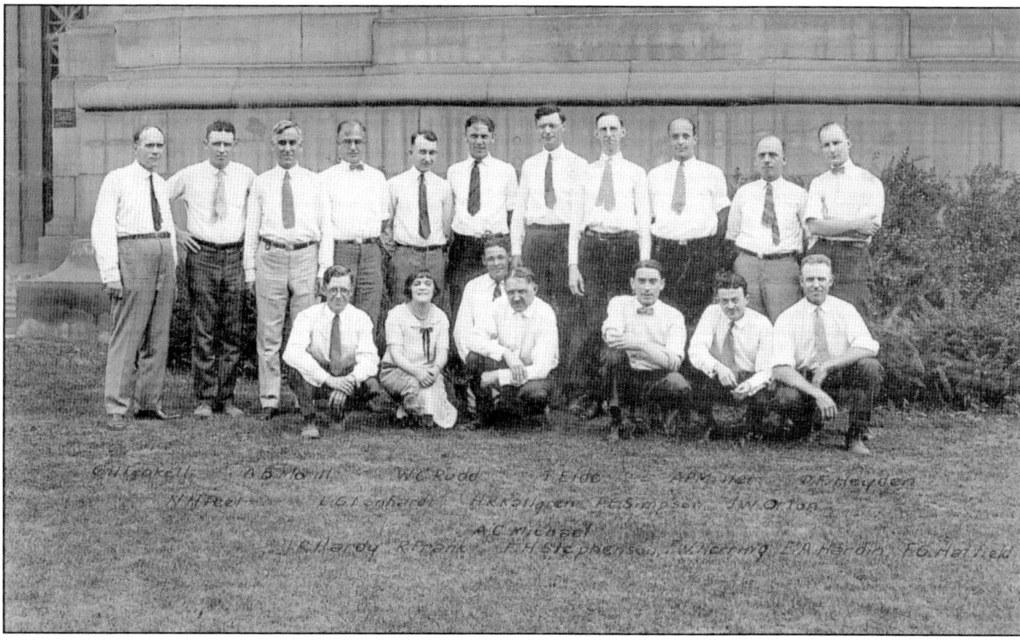

General superintendent of the Detroit Department of Water Services George Harrison Fenkell (top left) poses with members of his engineering staff in 1925. It was the responsibility of this group to find ways to squeeze ever more water from the water treatment plant that was already operating at maximum capacity during the booming 1920s. (DWSD.)

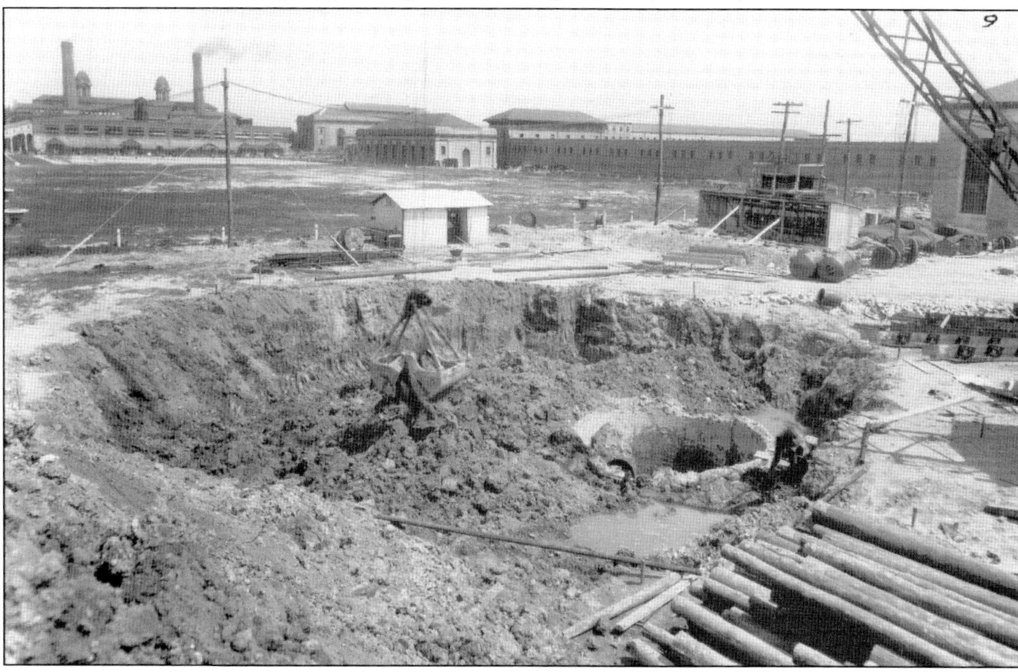

In June 1930, construction began near the foot of Water Works Park on a raw water tunnel to the new Springwells water treatment plant. The tunnel beginning here was eventually connected to the big water intake that had been installed in the Detroit River at the head of Belle Isle in 1932. (DWSD.)

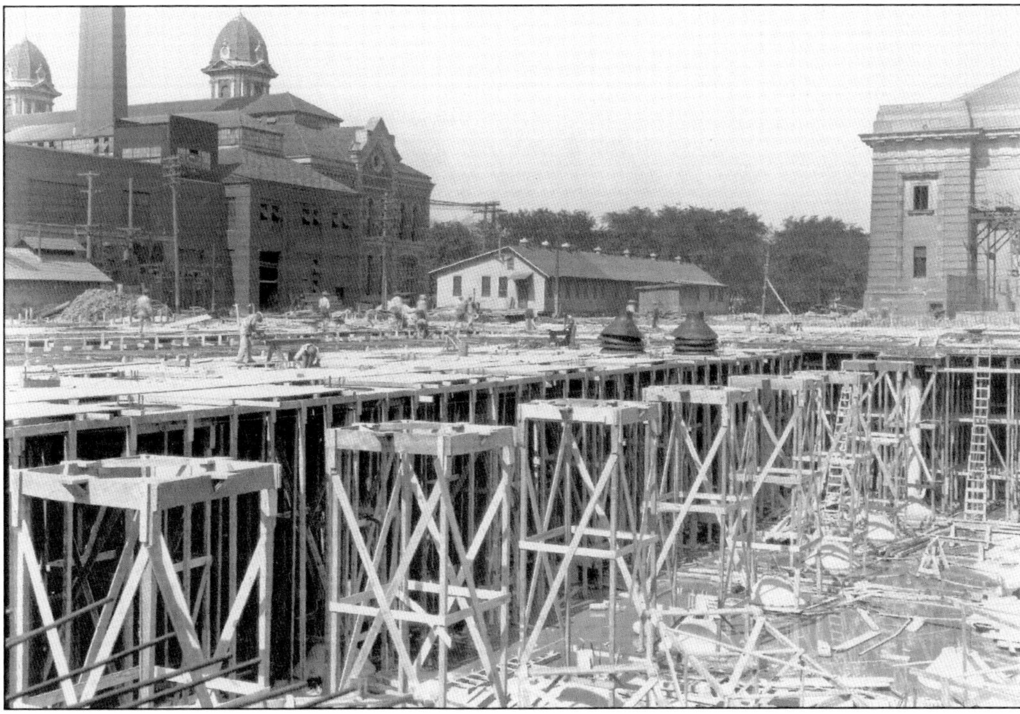

Reservoir No. 2, shown here behind Pumping Station No. 2, neared the end of construction on September 17, 1927. The below image shows the completed reservoir from the inside two months later, on November 19. This reservoir was just one of several stopgap measures employed in Detroit to extract additional water from the plant. Water reservoirs are filled during periods of low demand, usually at night. Water from the reservoir is then released into the water distribution mains to maintain pressure and meet higher demand throughout the day. (Both, DWSD.)

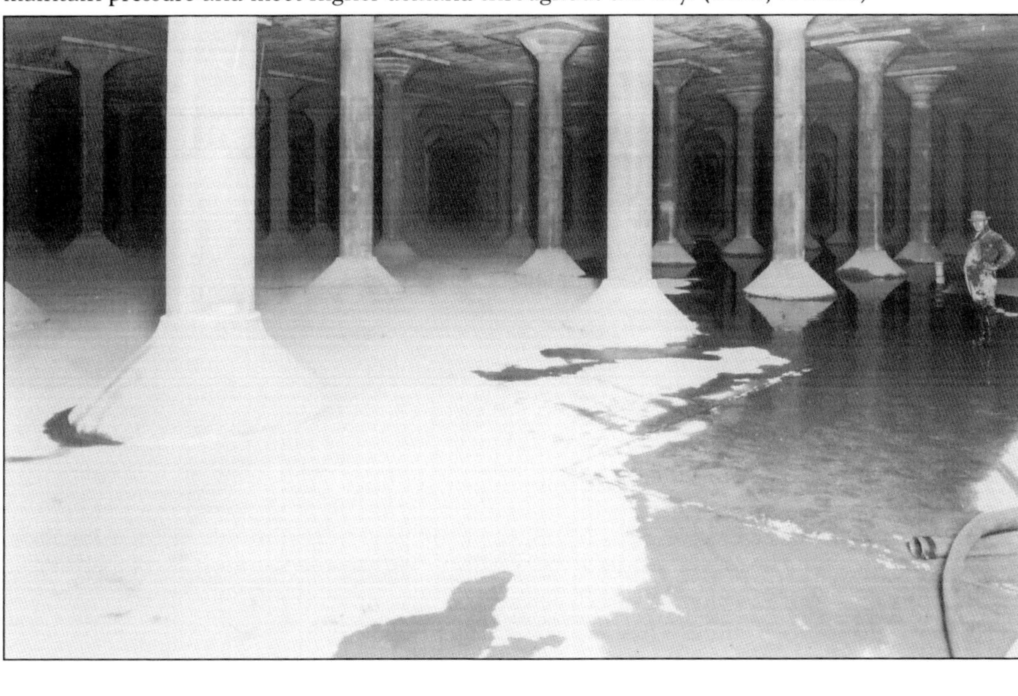

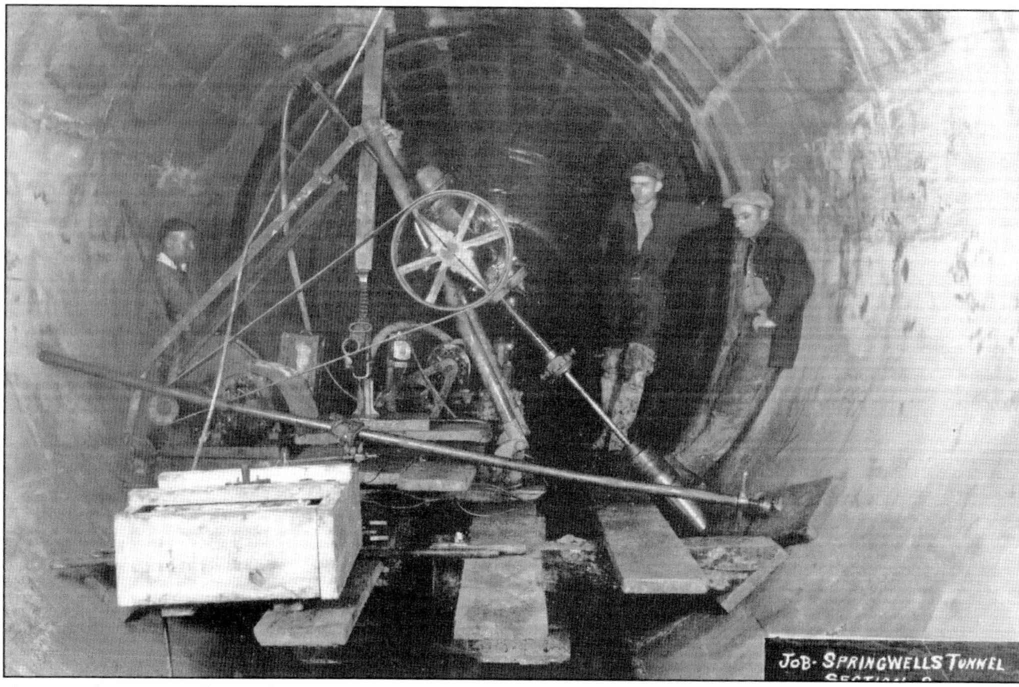

A tunnel crew works with an early tunnel-boring machine as it makes its way to the site of the Springwells plant on May 4, 1928. Some 60 years later, larger, more modern versions of this simple rig bored out the Chunnel beneath the English Channel between Britain and France. (DWSD.)

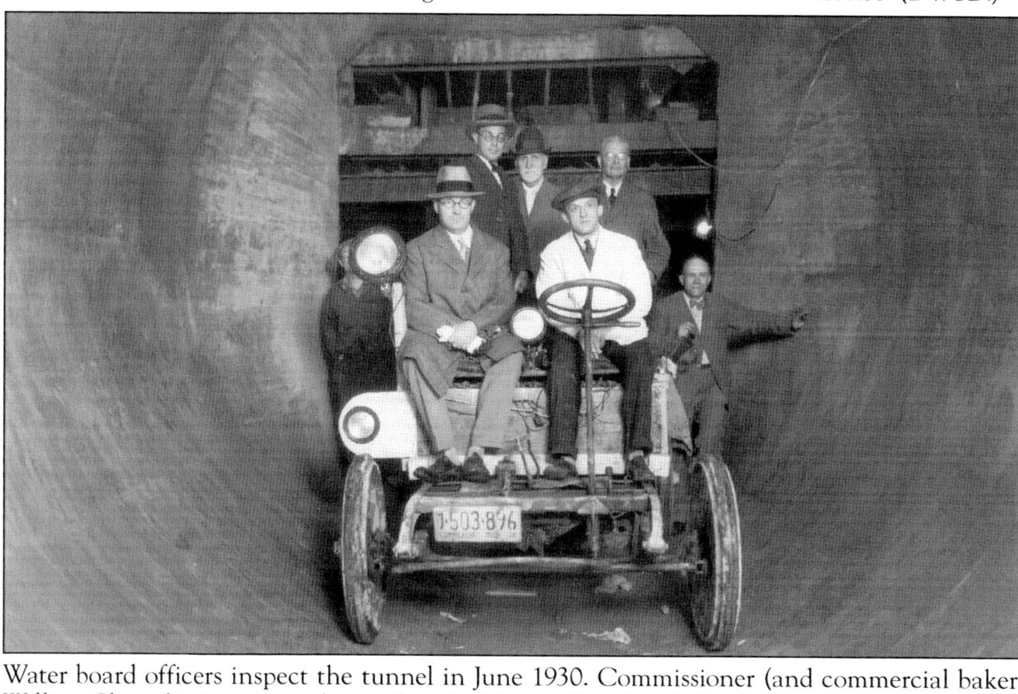

Water board officers inspect the tunnel in June 1930. Commissioner (and commercial baker) William Skryzcki sits next to the unidentified driver. Standing behind Skryzcki is future general superintendent Laurence Lenhardt. To his right are Alex Dow, president of Detroit Edison, and board secretary Daniel Grobbel. Standing at the far right is George Fenkell. (DWSD.)

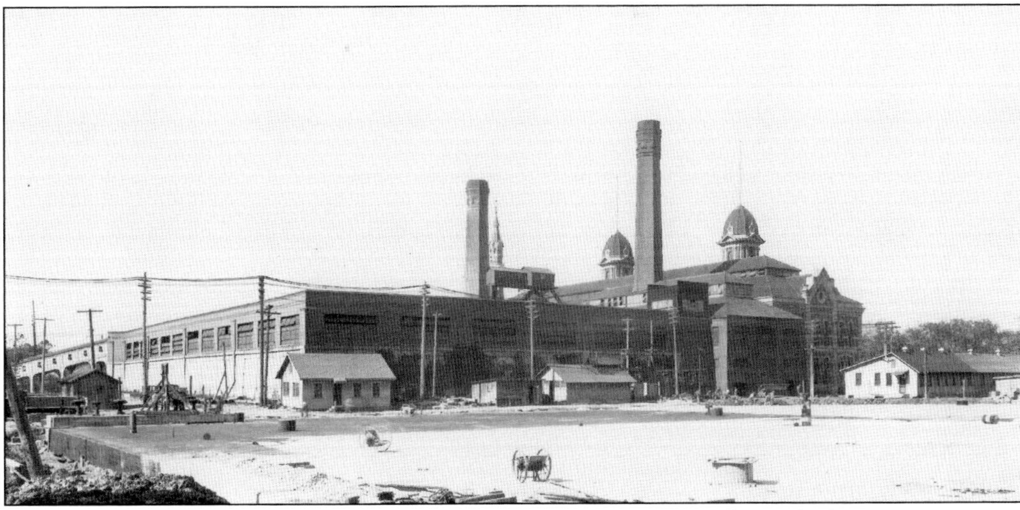

This photograph looks northwest from Reservoir No. 2 as it nears the end of construction prior to being covered over with soil in October 1927. The small object in the foreground is a manhole access. The larger building to the right is the coal shed and boiler house behind Pumping Station No. 1. The smaller structures are temporary construction offices. (DWSD.)

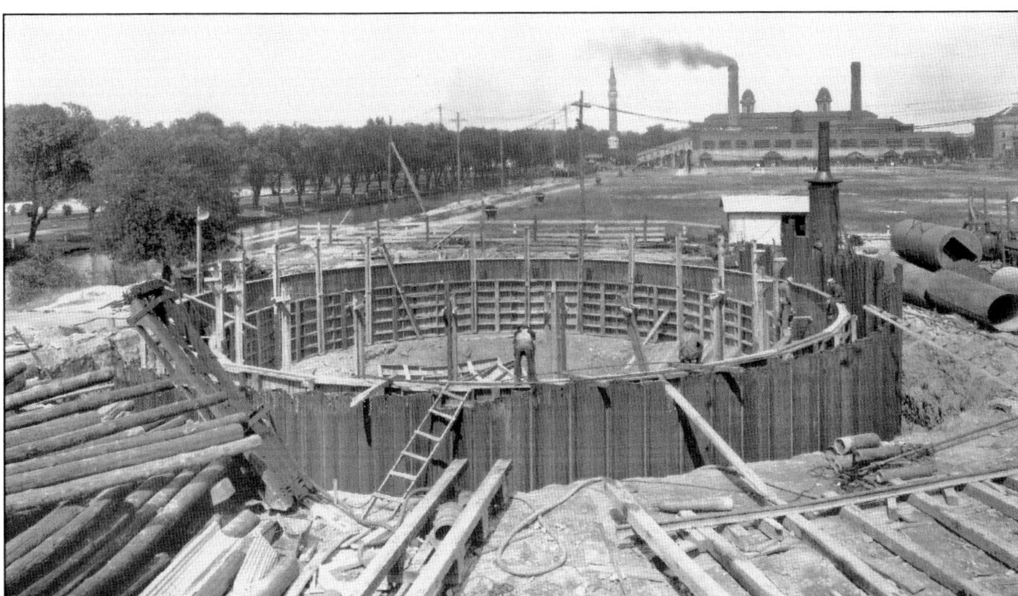

Construction of the low-lift pump at the foot of the Detroit River is ongoing in September 1930. The low-lift pump feeds chlorinated water by tunnel to the water treatment plants in Dearborn and northeast Detroit. A coaling vessel can be seen unloading its cargo by conveyor to the two pumping stations. The water tower and pavilion are at the head of the coaling canal. (DWSD.)

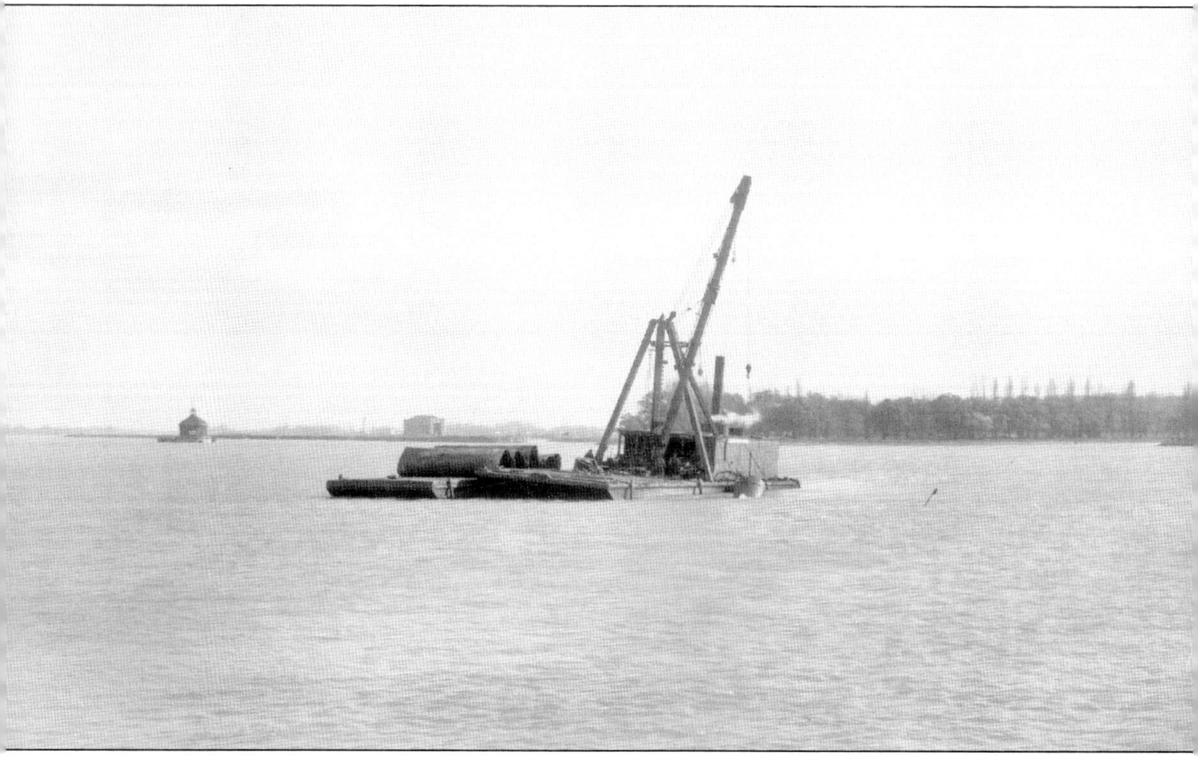

This view of the Detroit River looks south from Water Works Park in 1934. Beyond the dredging barge are Detroit's two raw-water intakes at the head of Belle Isle (the land mass at the right) The round structure at the left is the 1905 intake. The more elongated structure to its right is the 1932 intake. Canada can be seen in the extreme distance. (DWSD.)

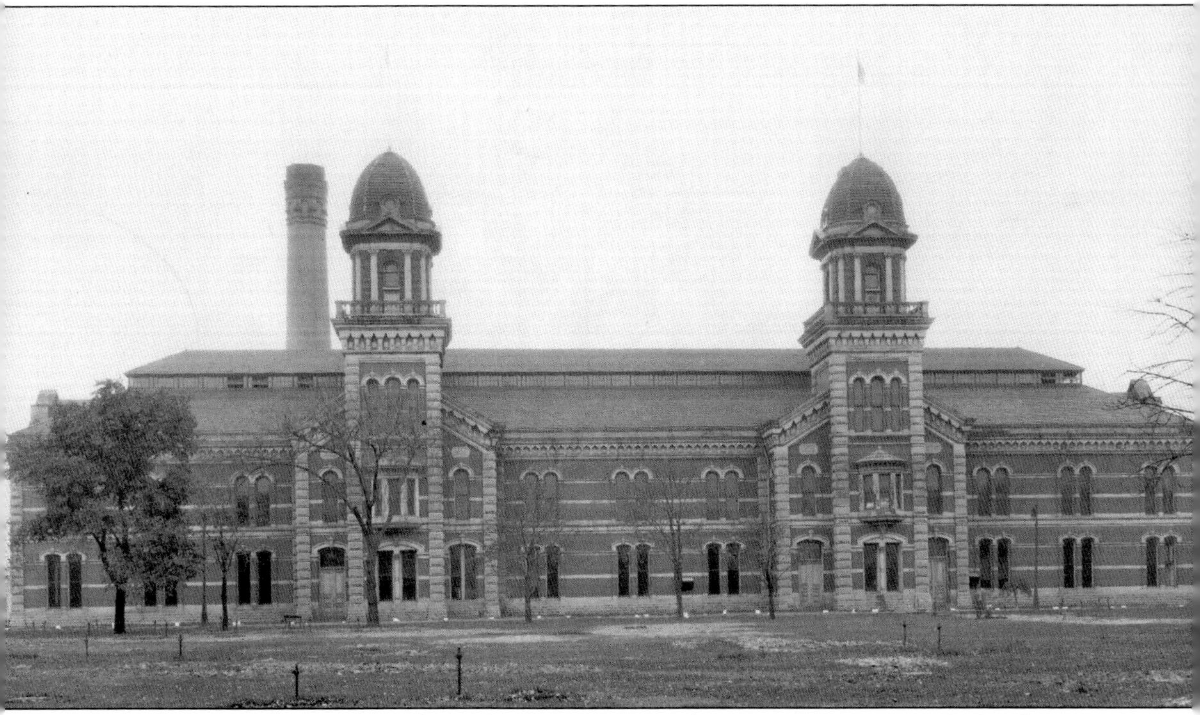

The original pumping station is shown here in 1936. It underwent expansions in 1886 and again in 1900. Expansion was necessary to make room for more steam-driven pumping engines needed to feed the unquenchable thirst of Detroit's rapidly growing industrial might. (DWSD.)

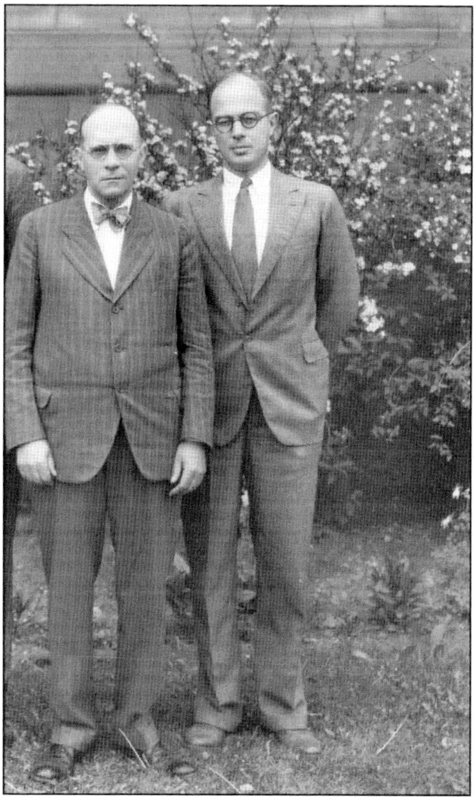

George Fenkell (left) and Laurence Lenhardt pose at Water Works Park in June 1931. Following his retirement in 1936, Fenkell's expansionist policies withered amid the austerity of the Great Depression. His successor, Lenhardt was decidedly more conservative. He told growing communities outside Detroit to develop their own systems rather than look to Detroit. Lenhardt planned to concentrate on maintaining the water infrastructure the city already had. (DWSD.)

The original pumping station is shown here in the 1950s, near the end of its long life. The twin-tower configuration and contemporary roof were added to the original building during its two major renovations. (DWSD.)

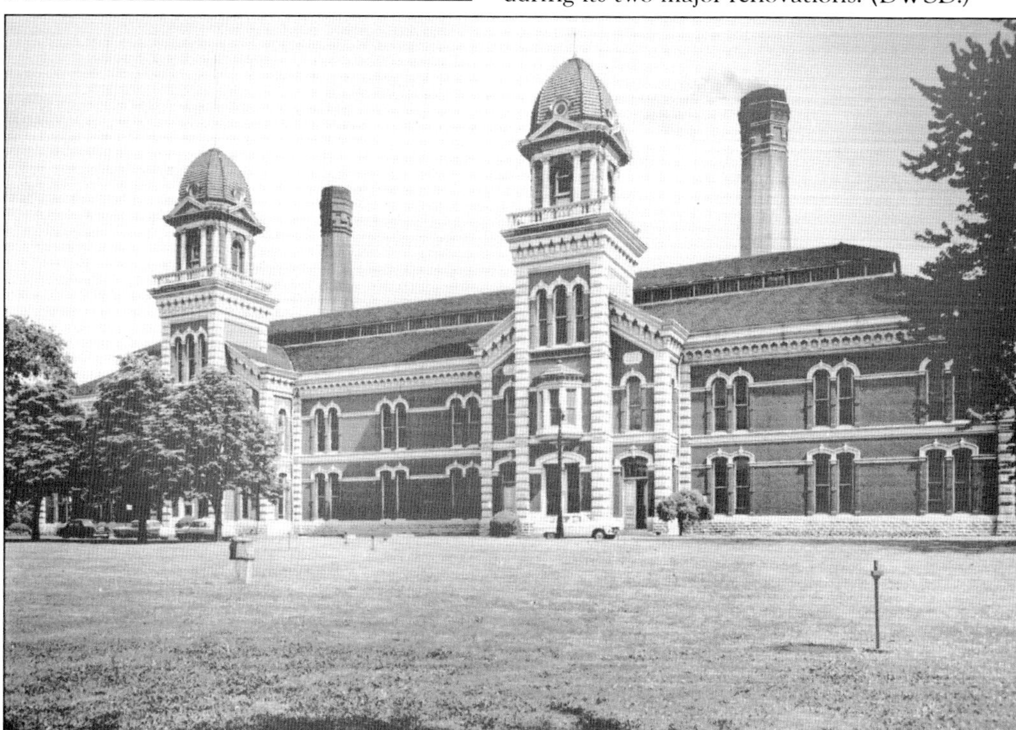

Four

Steam

Steam engines date to the first century of the Common Era, but they were mostly experimental curiosities, little more than toys. The aeolipile created by Hero of Alexandria and named for Aeolus, the Greek god of wind, was a metal sphere with nozzles on opposite sides. Water in the sphere was heated to boiling and then vented, causing the sphere to rotate. Had the Greeks recognized steam's potential, the industrial revolution that arrived in the 18th century might have occurred hundreds of years earlier.

The first practical steam engine was a water pump created by English inventor Thomas Savery in 1698, followed shortly by the first successful steam engines in 1712. These were "atmospheric" engines, using steam to create a partial vacuum, and were primarily used to pump water from coal mines. These engines powered the Industrial Revolution and remained the dominant source of power into the 20th century.

In the United States, steam has been used for a multitude of purposes, powering factories, trains, ships, and even early automobiles like the Stanley Steamer. It has been used to warm homes and offices and continues to be a major source of the electricity generated by the nation's power plants.

The first use of steam-driven pumping engines in a drinking water system occurred in Philadelphia in 1799. Steam was first used to power pumps in Detroit's water distribution system in 1830, utilizing a 10-horsepower engine shared with a foundry located near the present site of Cobo Center. From there, the engines grew larger and more powerful. The engine installed in Water Works Park in 1886 served for more than half a century and was widely believed to be the world's largest steam engine for most of its life.

Steam engines were big, soaring 30 feet or higher into the air and an equal distance belowground. They were quiet, and they were powerful, but they were also inefficient and dirty, because of the coal powering them. It all came to an end in 1962 when Water Works Park switched over to more powerful and fuel-efficient electric pumps.

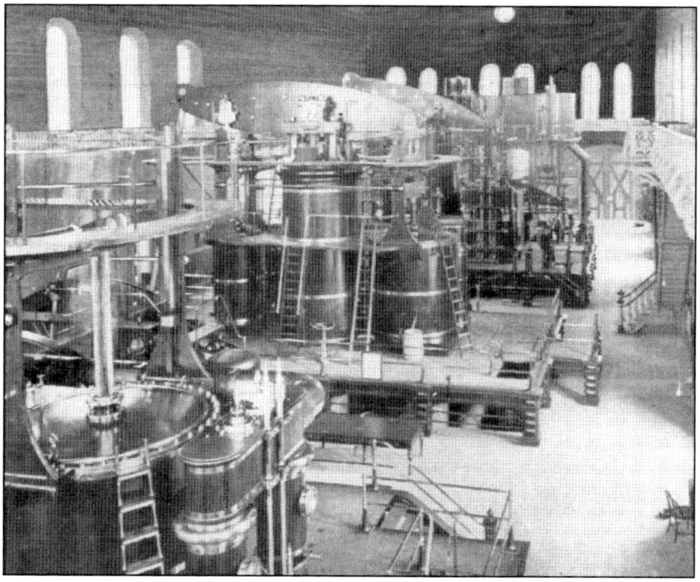

This undated image shows the steam-driven pumping engines inside Pumping Station No. 1. This station, in its various different configurations, was the sole source of drinking water in Detroit from 1879 to 1914, when Pumping Station No. 2—now the High Lift Building—came on line. (DWSD.)

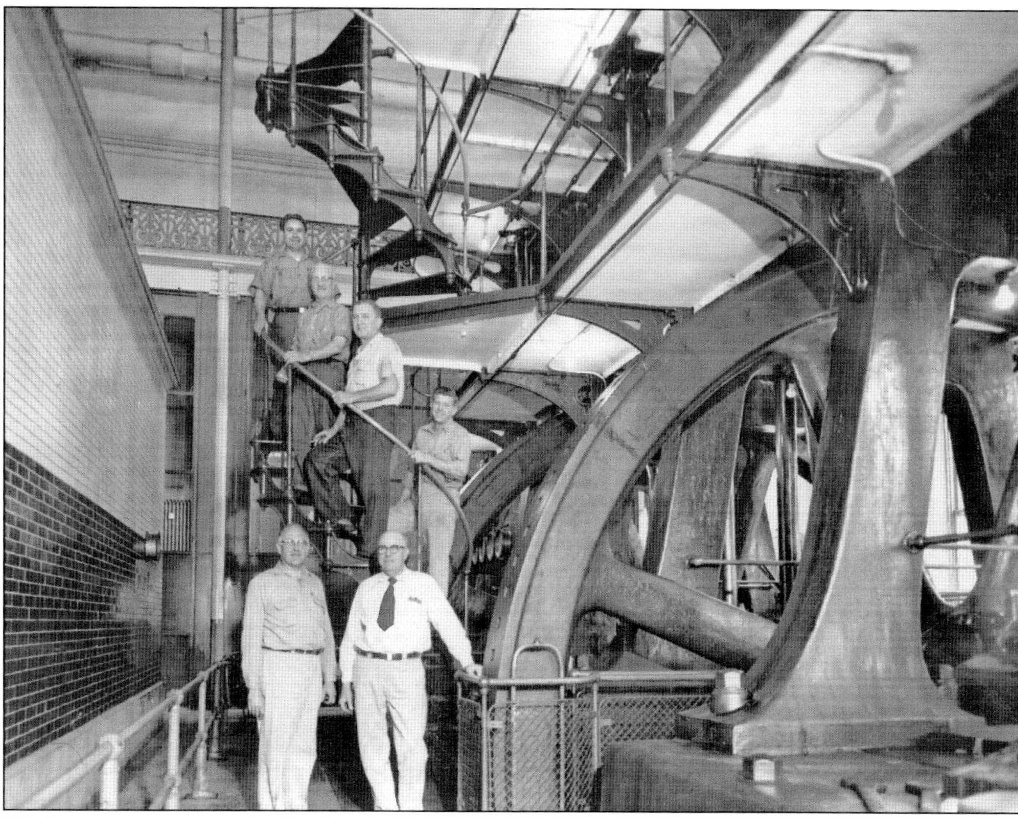

Workers pose next to the flywheel of one of the big steam engines at Water Works Park in the mid-1950s. A single revolution of each flywheel translated into nearly 1,300 gallons of water pumped. Until they were joined by those at Detroit's Springwells station in 1935, the engines at Water Works Park were responsible for pumping the entire drinking water supply for all of Detroit's residential and industrial users. (DWSD.)

This image of the original pumping station and water tower at Water Works Park, probably from around 1880, shows the pumping station in its original configuration, when it contained only a single steam engine. A second engine was installed in 1881. The station was enlarged in 1886 to accommodate a third engine, and expanded again in 1900. (Author.)

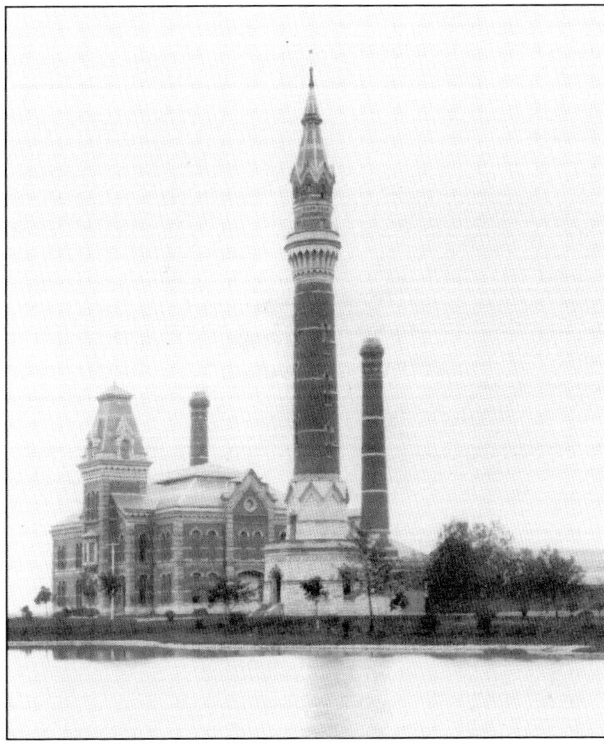

This image shows the original pumping station and water tower as they appeared after the 1886 expansion. The station, only seven years old at the time, needed to grow to make room for additional steam engines to supply increasing industrial demand. The engine installed that year, believed to be the world's largest at the time, served faithfully for more than 50 years. (DPC/LOC.)

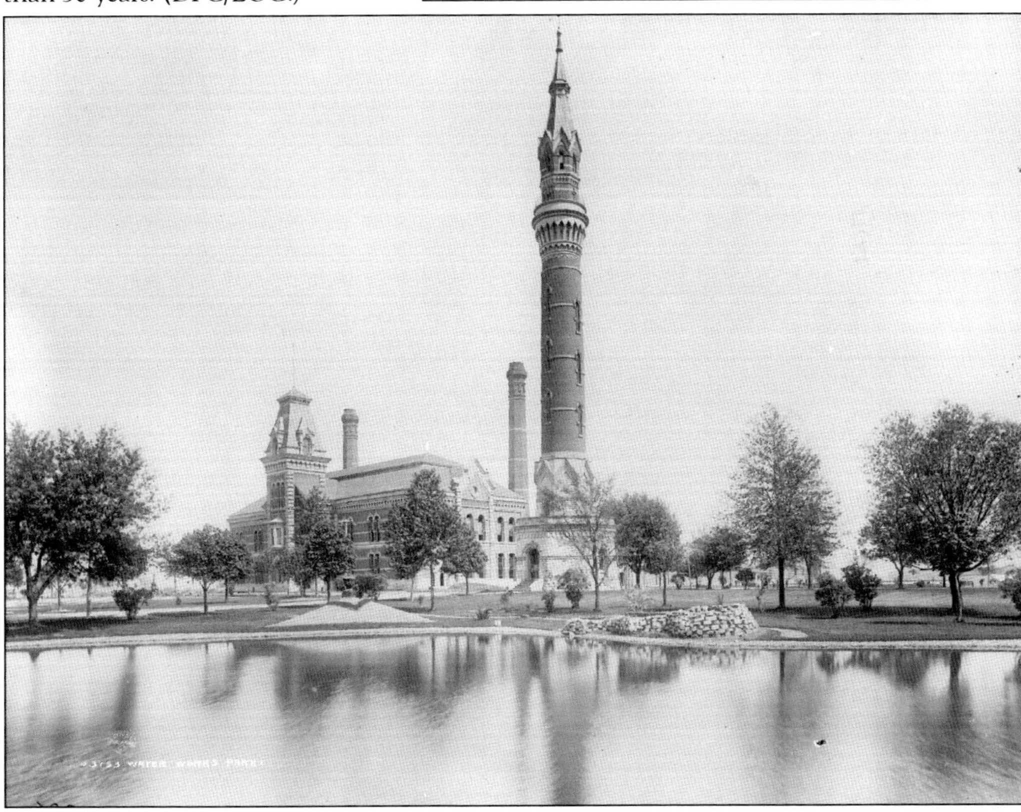

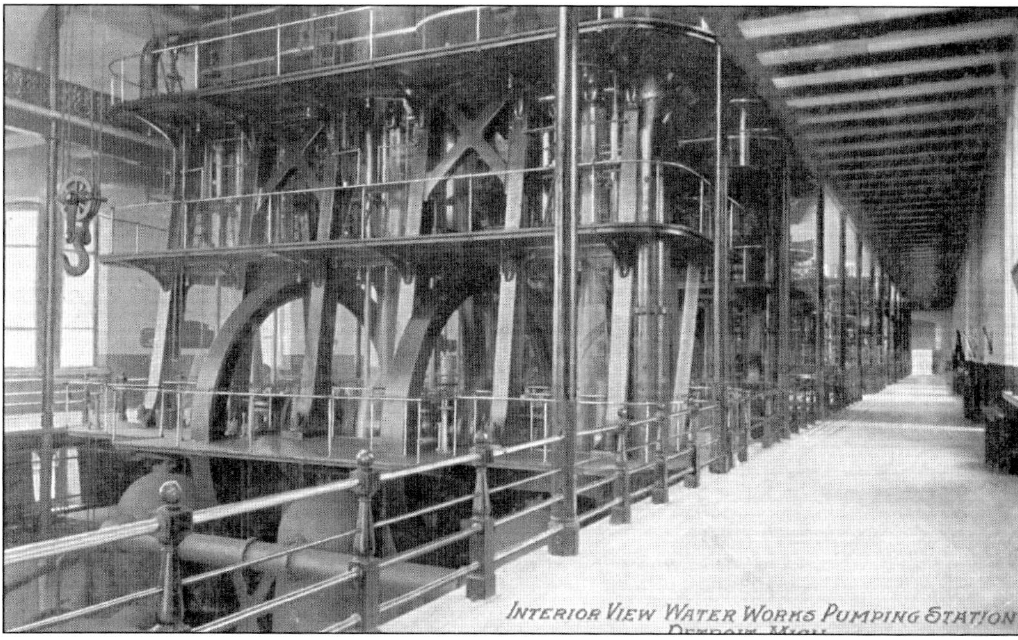

This hand-tinted souvenir postcard shows some of the massive steam engines at Water Works Park. These giants usually weighed several tons, and extended 30 feet into the air and 30 feet belowground. A single revolution of an engine's huge flywheel translated into well over a thousand gallons of water forced into the discharge mains that fed Detroit's water distribution system. (Author.)

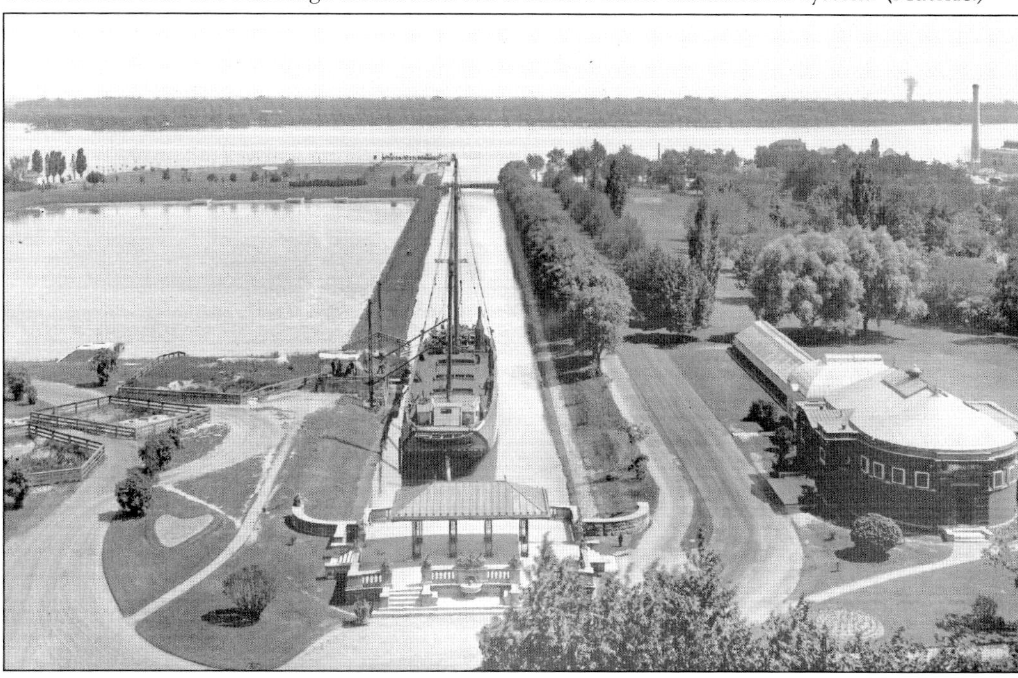

A coaling vessel unloads cargo at Water Works Park some time before 1920, as seen from the observation deck of the water tower. The Hurlbut Library and greenhouses are on the right. The pavilion, where bands entertained visitors on Sundays, is in the foreground. An open settling basin is to the left and the Detroit River and Belle Isle are in the distance. (DWSD.)

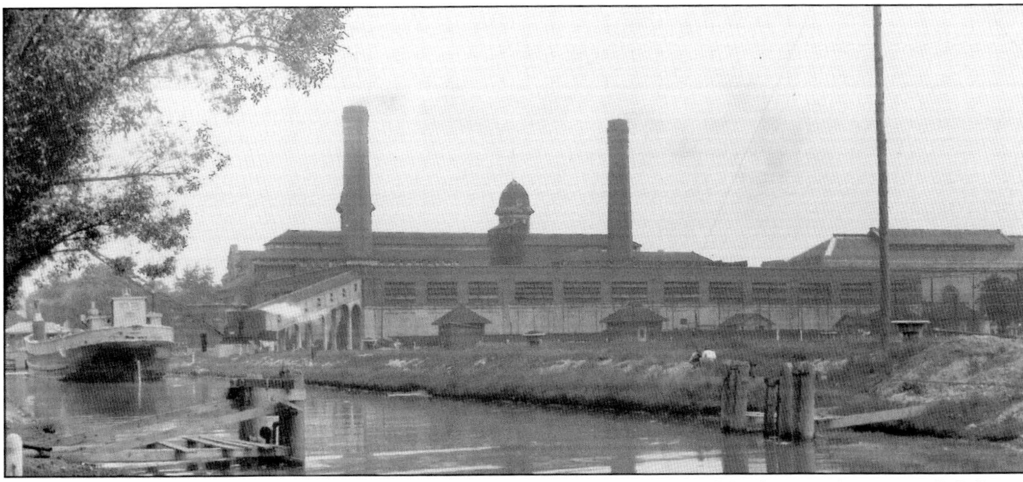

Seen here from the canal gate at the Detroit River on June 30, 1930, this coaling vessel delivers cargo. The back of Pumping Station No. 1 is shown here, with Pumping Station No. 2 to its right. The park was once home to 10 of these coal-fueled, steam-powered behemoths. (DWSD.)

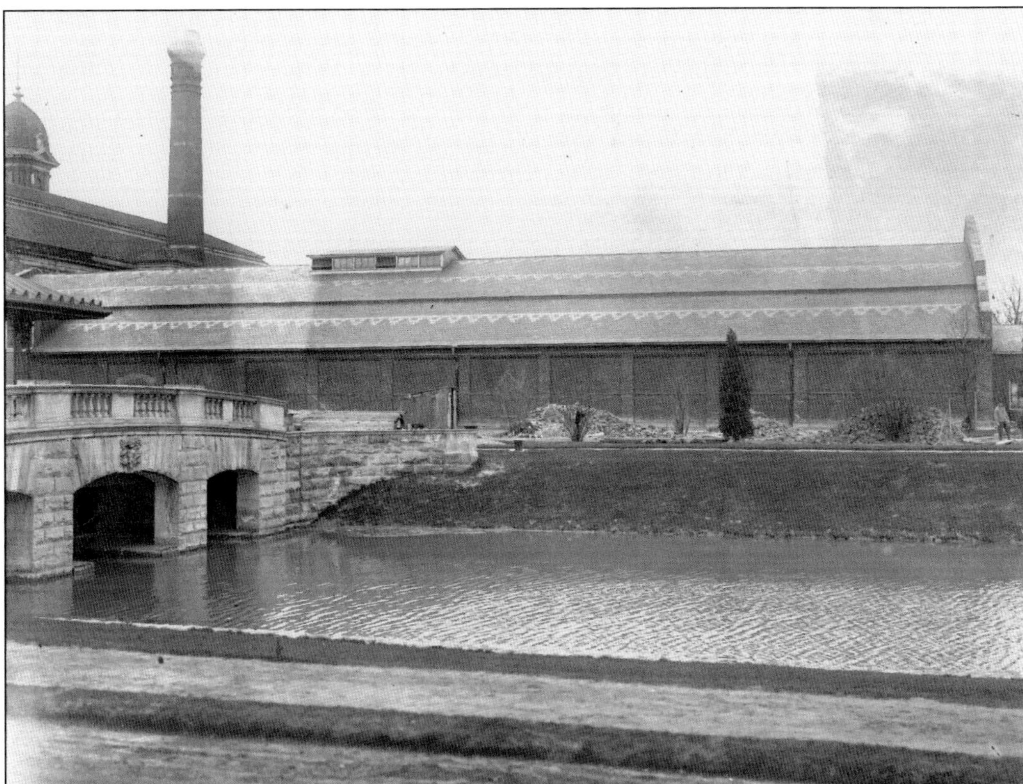

The coal shed is shown here near the pavilion at the north end of the coaling canal in 1908 or 1909. Each one of the big boilers feeding steam to the huge "walking beam" steam engines was capable of burning up to 11,000 pounds of coal per hour. (DWSD.)

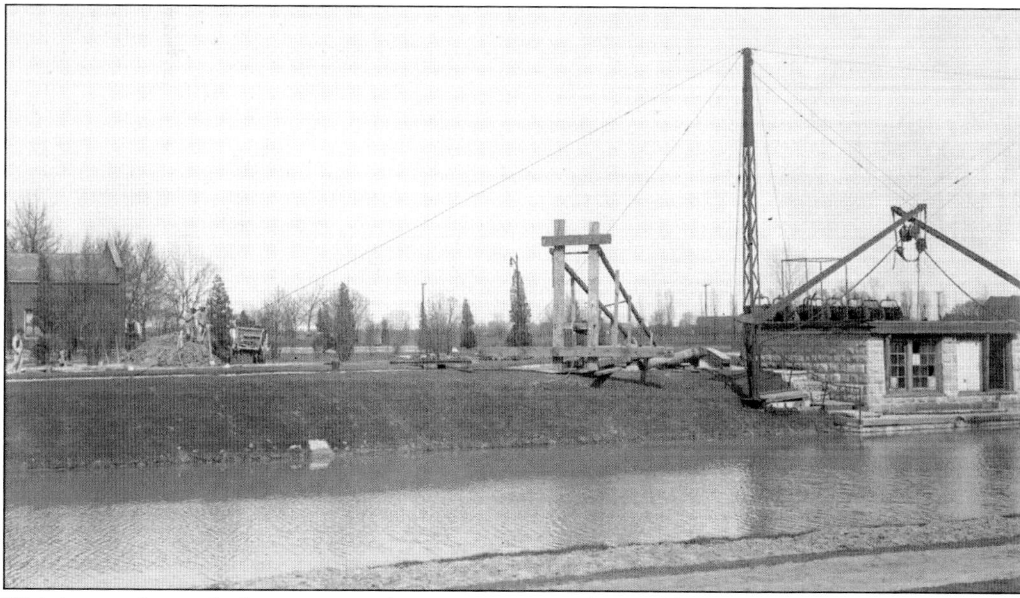

This image shows the beginnings of construction on the coal-handling bridge and conveyor at Water Works Park. The bridge and conveyor expedited the transfer of coal from coaling ships into the coal shed (to the left) during the park's coal-fired years. (DWSD.)

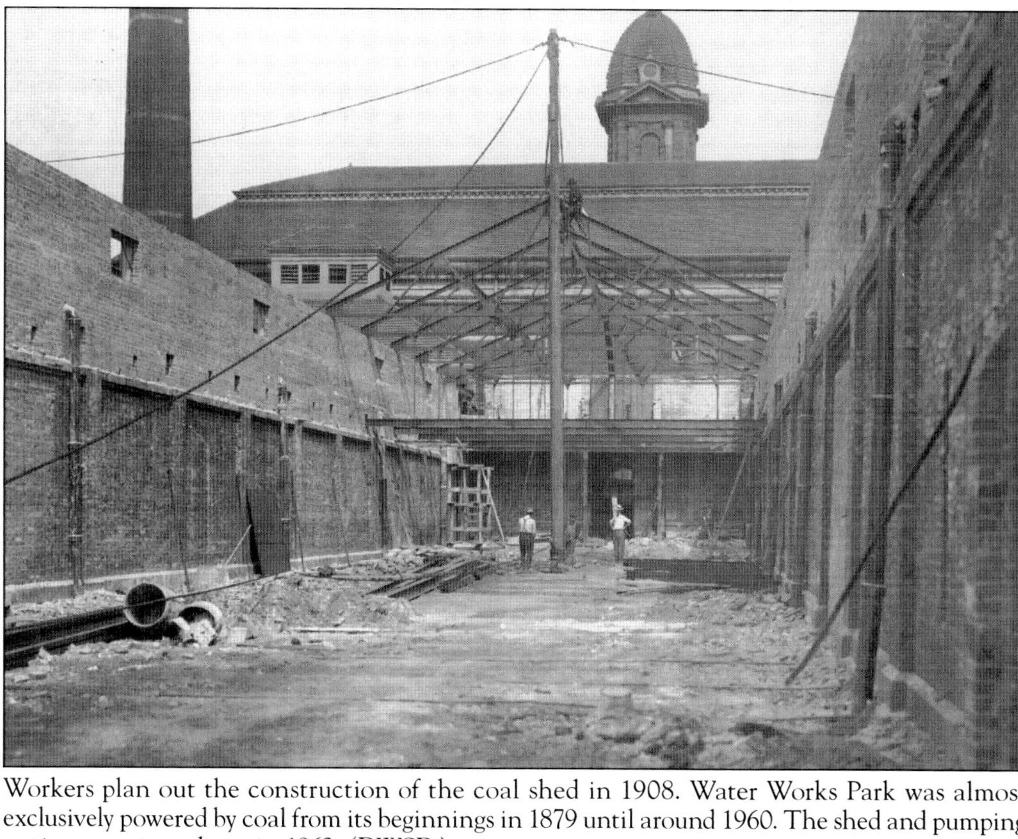

Workers plan out the construction of the coal shed in 1908. Water Works Park was almost exclusively powered by coal from its beginnings in 1879 until around 1960. The shed and pumping station were torn down in 1962. (DWSD.)

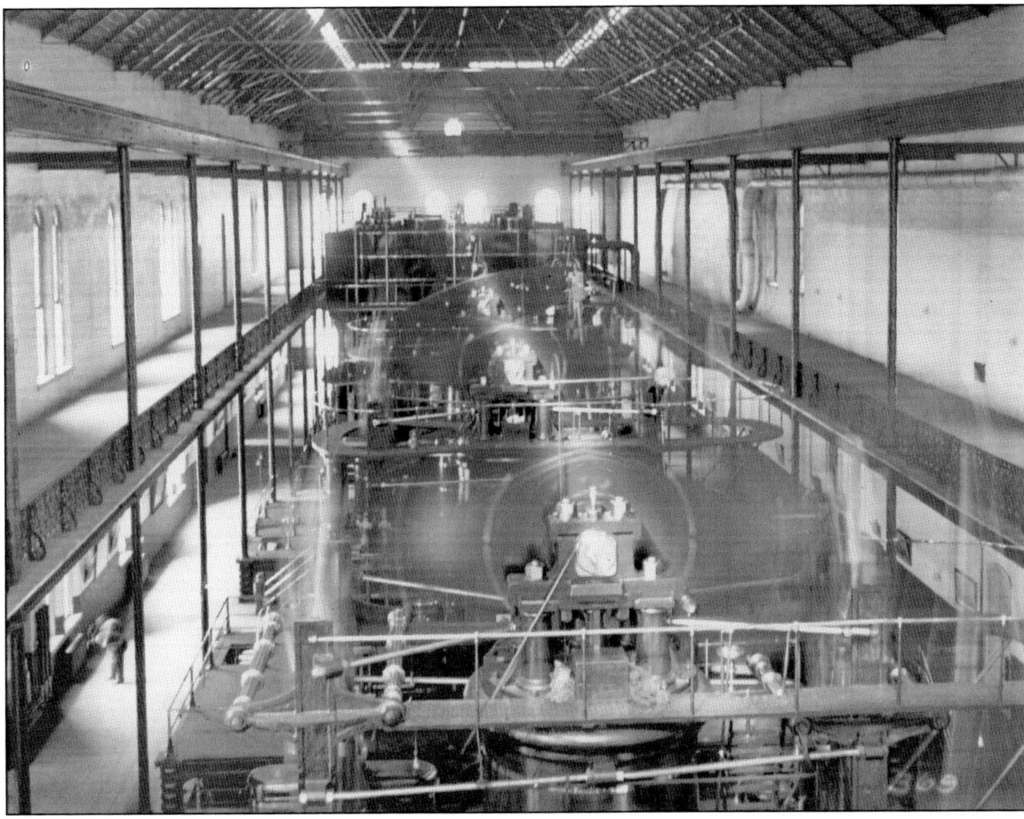

The big steam engines work hard inside Pumping Station No. 1 in this undated photograph. At one time, this building held six of the steel giants. The building was demolished in 1962 after the steam engines it housed were dismantled and removed. (DWSD.)

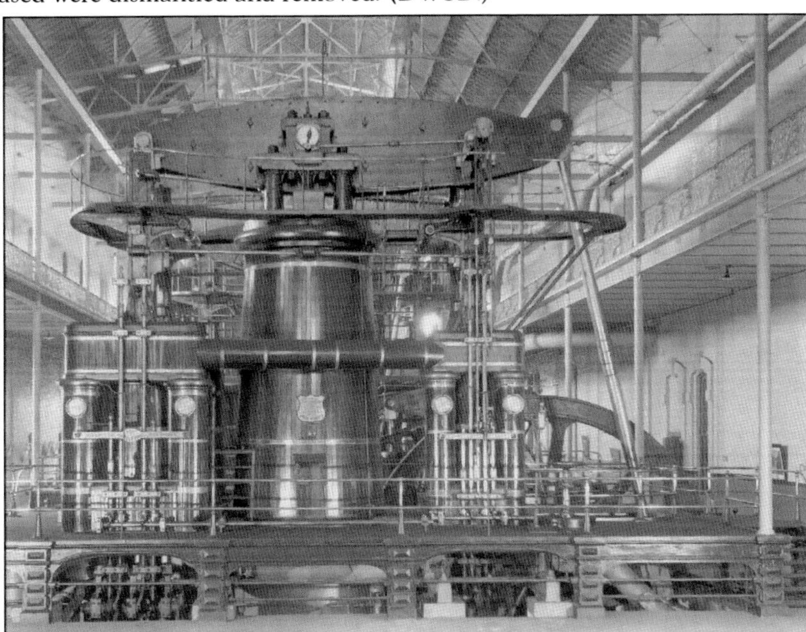

Engine No. 2 was a "walking beam" engine, so called because the engine's rocking resembled a walking motion. Walking beams were the most popular steam engines in the United States and were manufactured all the way until the 1940s. (DWSD.)

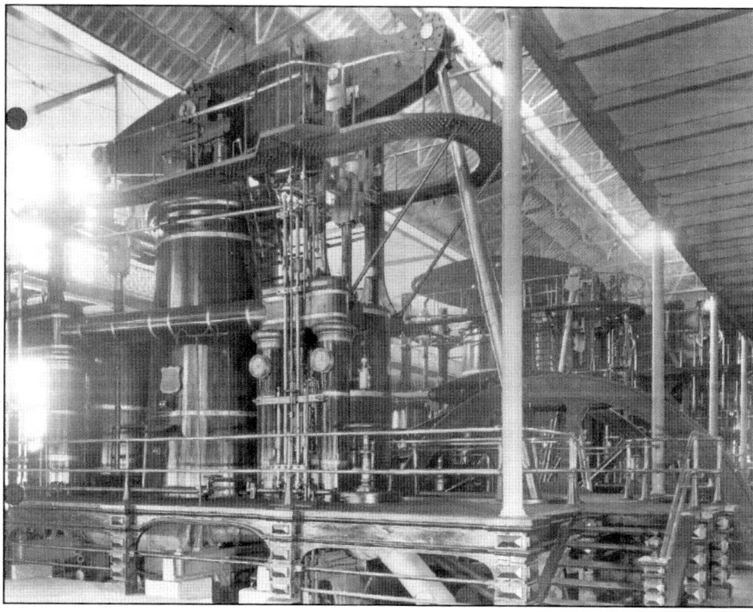

This undated photograph shows the ground level of the steam engines in Pumping Station No. One at Water Works Park. (DWSD.)

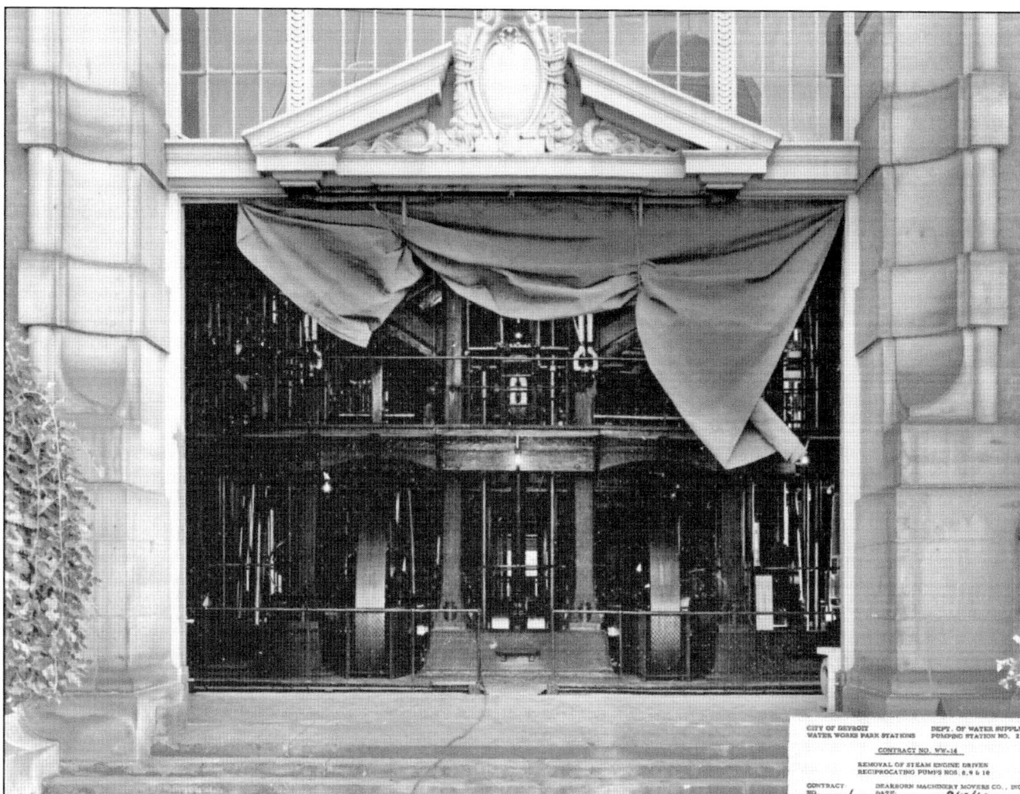

This is a last look at Pumping Station No. 2 on September 13, 1960, just prior to the dismantling and scrapping of the engines within. The engines were replaced by more powerful and fuel-efficient electric pumps as part of Detroit's plans to expand the size of its drinking water distribution system. (DWSD.)

Five

End of an Era

With the exception of its first pump—the horse-powered unit installed on a wharf at the foot of Randolph Street in 1824 by Peter Berthelet—Detroit's drinking water distribution system was steam-driven for more than a century. Giant engines supplied the water needs of Detroit's auto manufacturers during the birth of the Motor City around 1900. Steam, and a rapidly expanding water distribution system, supplied the needs of industry in the 1920s, when Detroit was the manufacturing capital of the country—a leader in airplane manufacturing, pharmaceuticals, steel, and more than 30 other industries, including the illegal importation of liquor from Canada.

The onset of the Great Depression in 1929 brought an abrupt end to the good times and plunged the city into an economic morass. Prosperity returned as Detroit assumed the role of Arsenal of Democracy in World War II, but vivid memories of hard times caused city leaders to reject further expansion of the water system.

But expansion returned with a vengeance in the 1950s as baby boomers and returning GIs used Dwight Eisenhower's new system of freeways to move from Detroit to the suburbs. It was cheaper for these undeveloped suburban communities to hook up to Detroit's already established distribution system rather than invest the enormous sums of money necessary to build their own systems from scratch.

To facilitate system expansion, the decision was made to electrify Water Works Park and replace the old steam giants with smaller, more powerful, and more fuel-efficient electric pumps. Once this was accomplished, there was no looking back, as Detroit's system grew in the 1950s and 1960s and all the way into the 21st century. Currently, Detroit's water treatment system is the third-largest in the United States. Its five water treatment plants service 43 percent of Michigan's population in an area of 1,079 square miles, including Detroit and 126 other communities in southeast Michigan.

Efforts were made to sell the park's old steam engines, but with no buyers, they were dismantled and cut up for scrap.

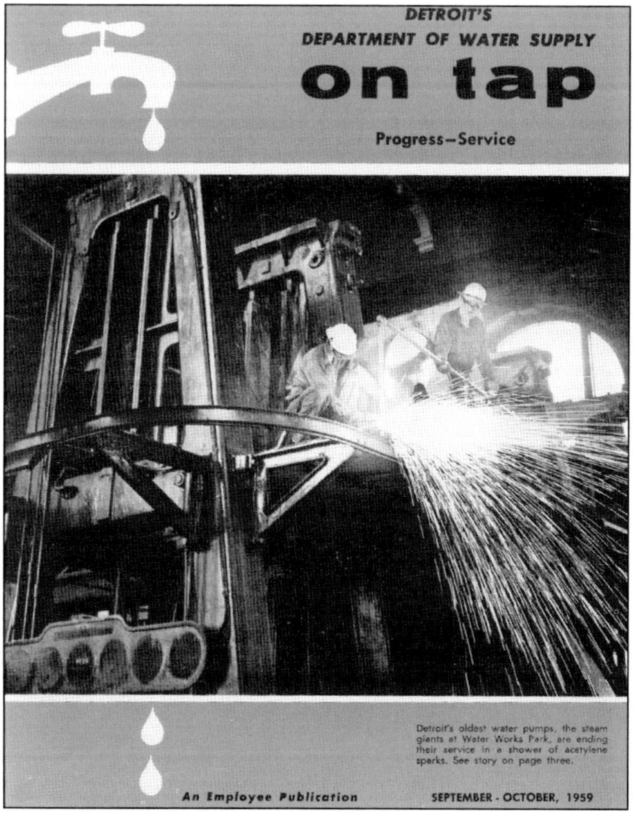

Electrification, and the dismantling of the steam giants, is the story featured on the cover of the Detroit Water Department employee newsletter for September-October 1959. "For the next several months riggers will be climbing about the three story tall machines," employees are informed. "With acetylene torches and sledge hammers they will cut the units apart piece by piece." (Author.)

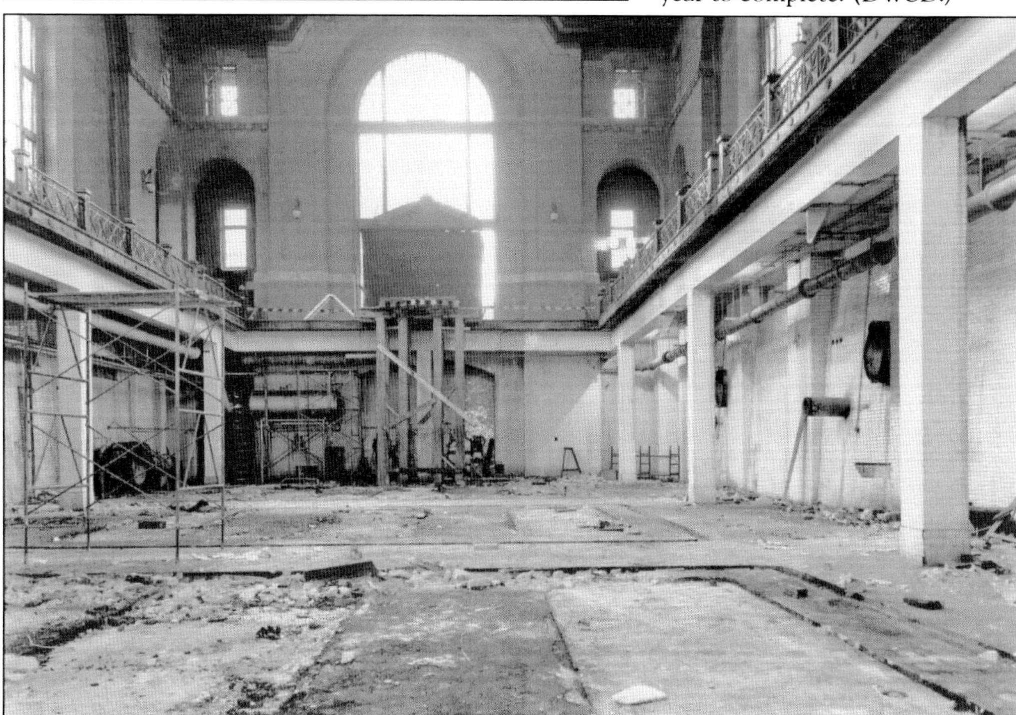

Devoid of the steel behemoths it once held, Pumping Station No. 2 seems especially empty and cavernous in November 1960. Dismantling and removal took more than a year to complete. (DWSD.)

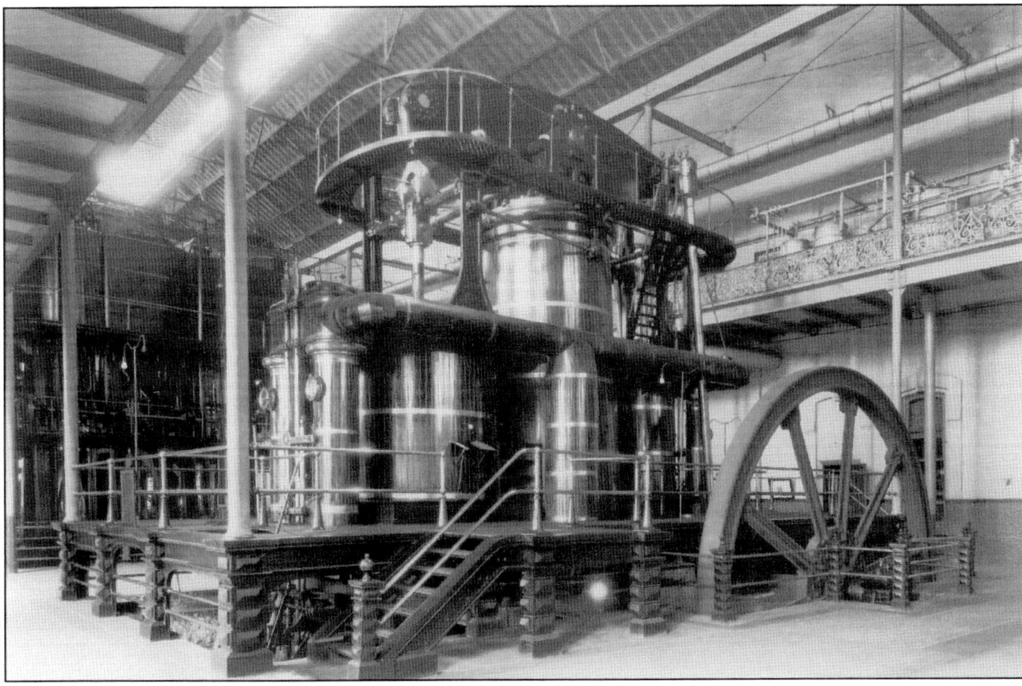

The above photograph, from October 14, 1927, shows just one of the many steam-driven, "walking beam" pumping engines made of gleaming steel. More than 50 years later, it is indistinguishable from the below image of Engine 12 in Pumping Station No. 2, from October 1959. When the original pumping station went on line in 1879, the entire output of Water Works Park was supplied by a single steam-driven pump. By 1959, when it was electrified, there were 10 huge engines, and Water Works Park was just one of three water treatment plants supplying Detroit's water. (Both, DWSD.)

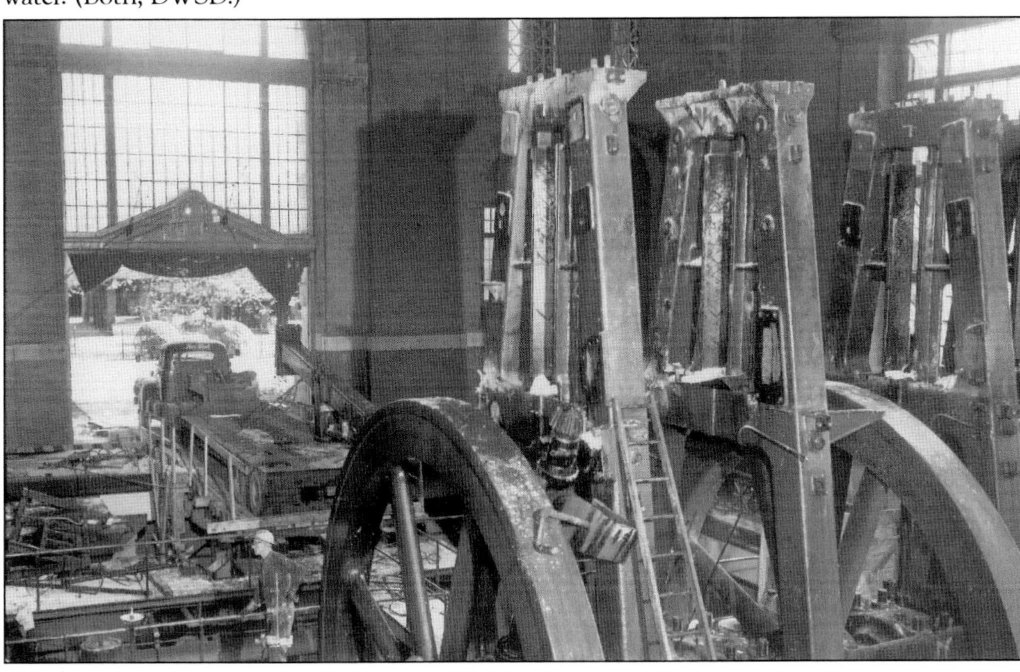

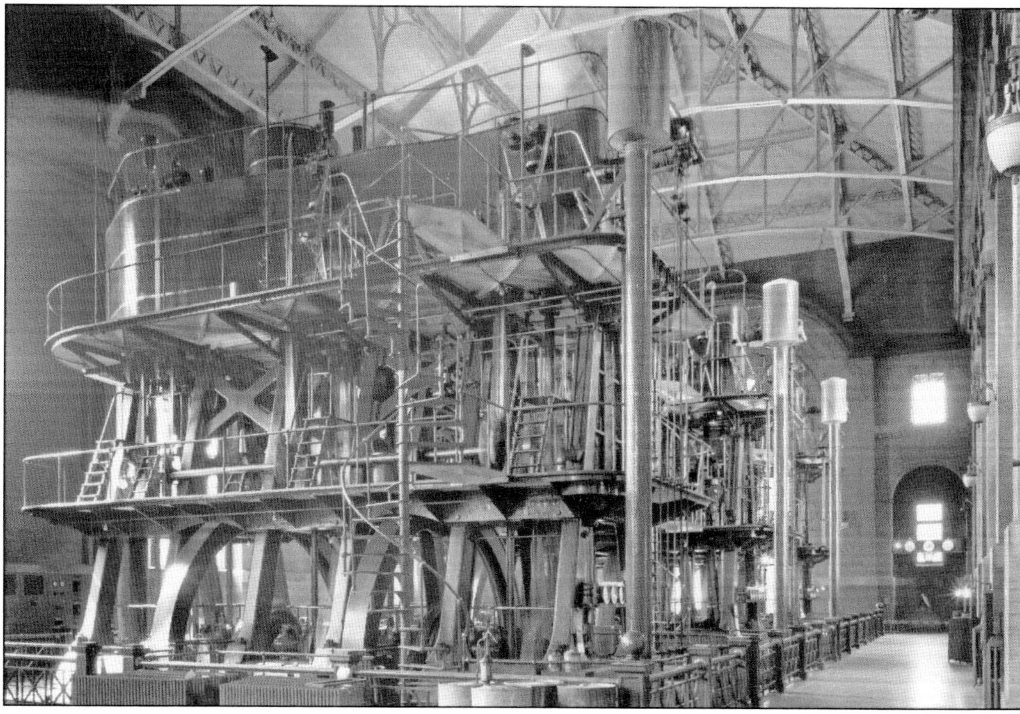

Pumps 8, 9, and 10 are shown here on September 13, 1960, in their final days in Pumping Station No. 2. (DWSD.)

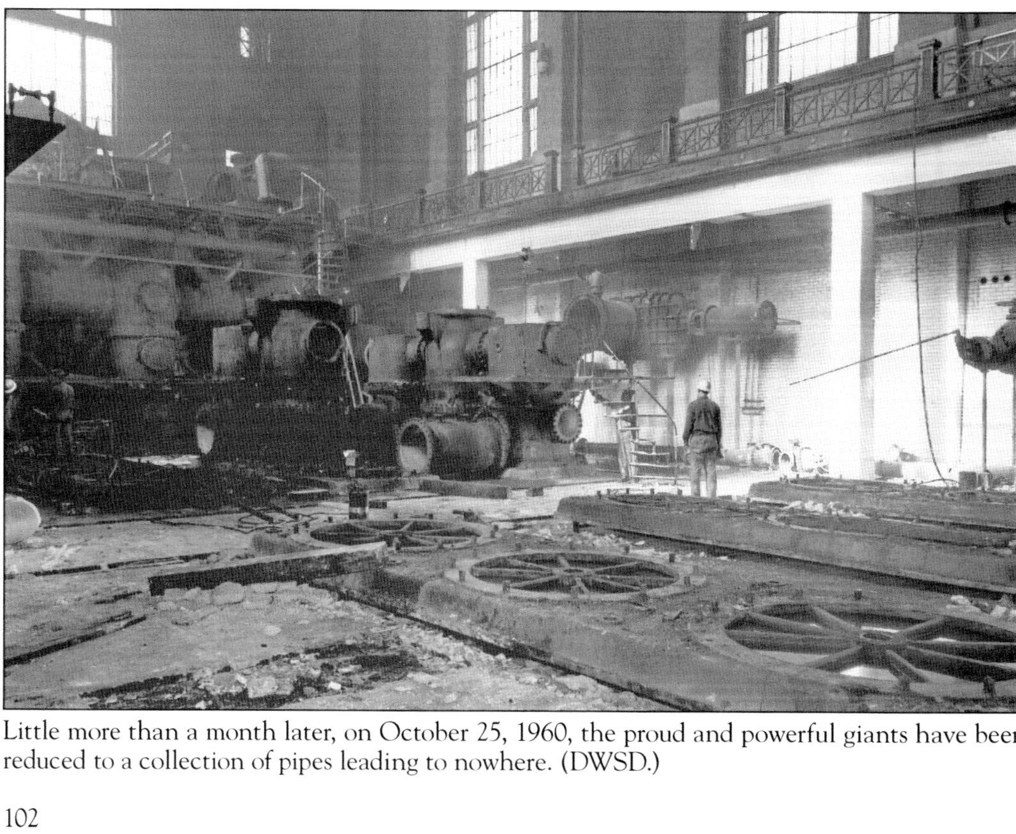

Little more than a month later, on October 25, 1960, the proud and powerful giants have been reduced to a collection of pipes leading to nowhere. (DWSD.)

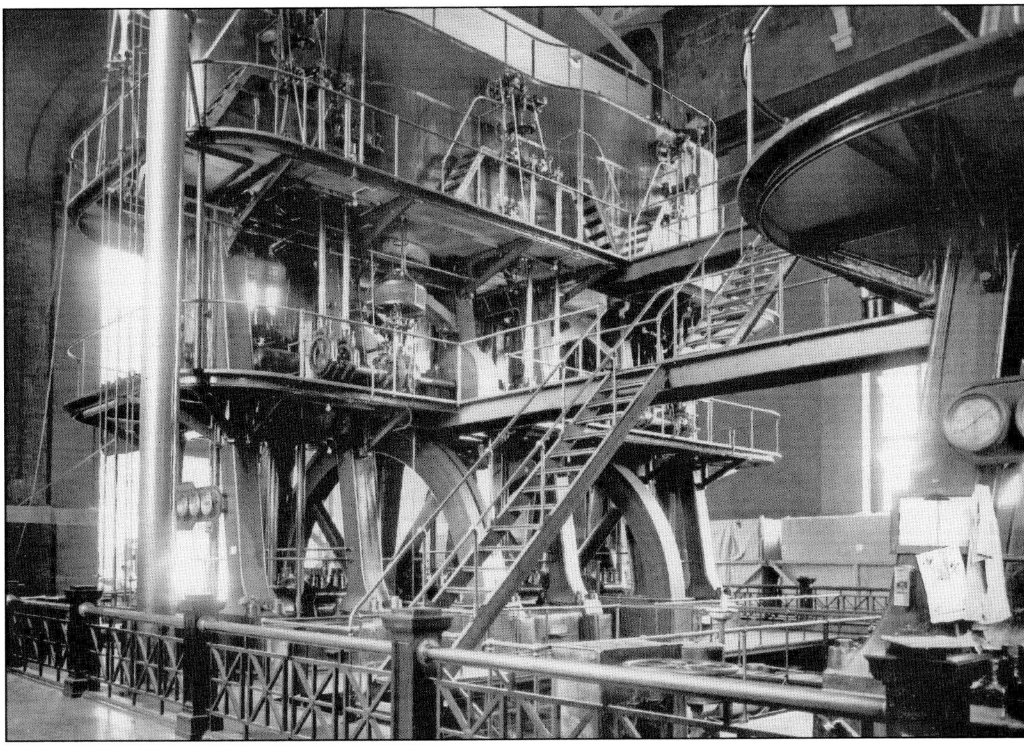

Venerable steam-powered Engine 13 is shown here only 11 days apart—above on September 2, 1959, and below on September 13, 1959, after the removal of its flywheels and A-frames. The truck is waiting to be filled with once valuable metal, now just scrap. (Both, DWSD.)

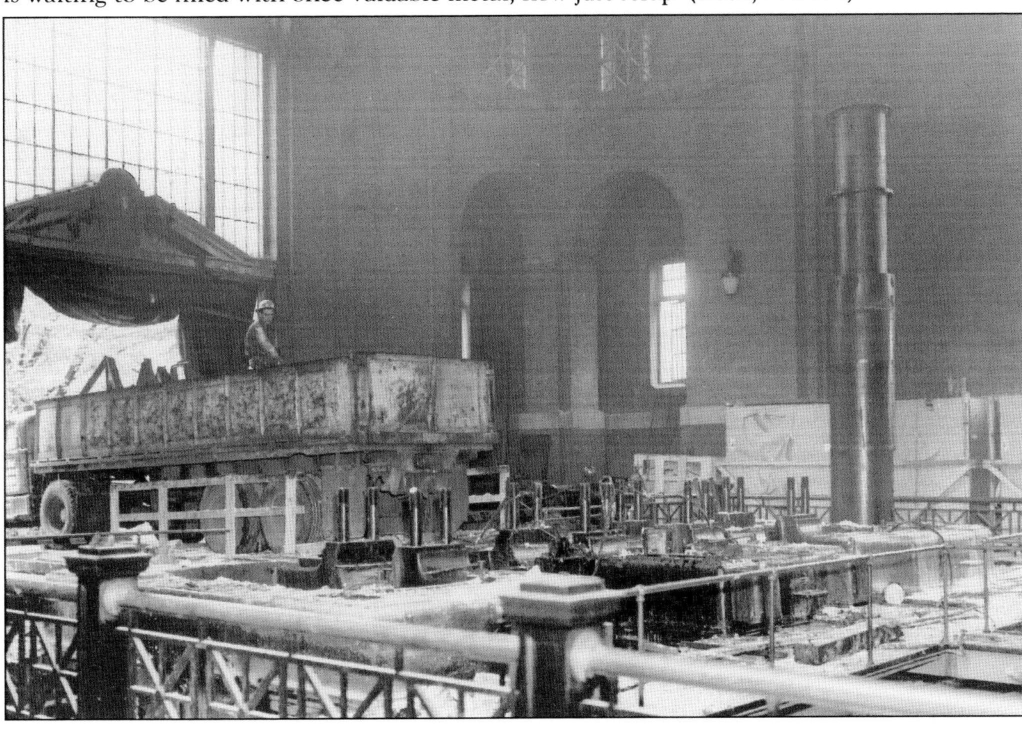

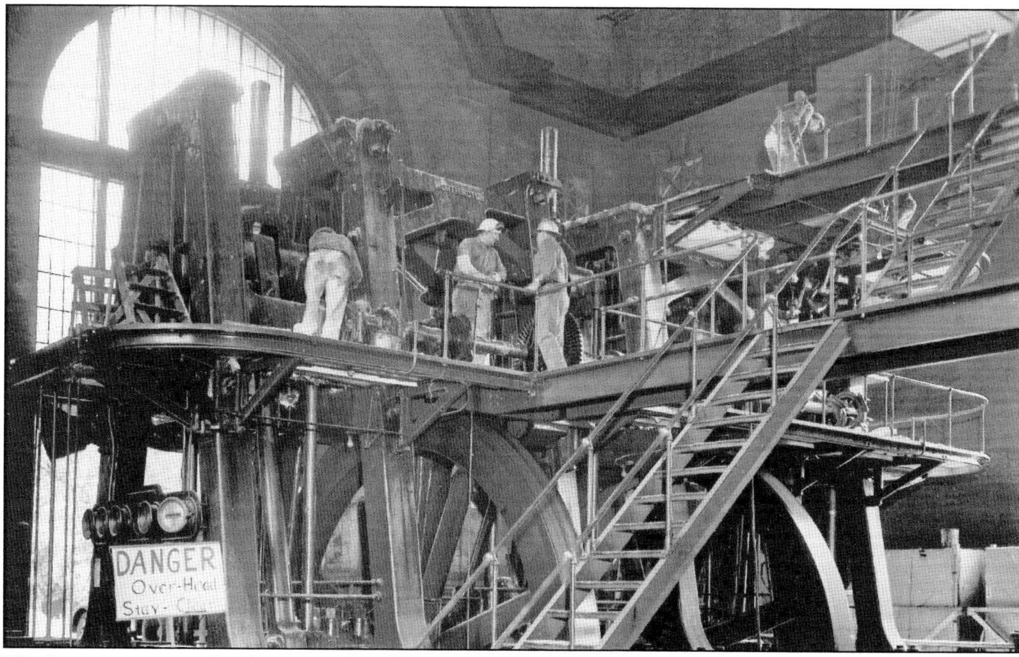

Workers engage in the disassembly of Engine 13 in the High Lift Building on September 16, 1959. Dismantling the old steam engines was a dirty and labor-intensive job. Once everything that could be was removed, the main support structures and other large sections were cut into smaller pieces by workers with acetylene torches; the smaller sections were disposed of. (DWSD.)

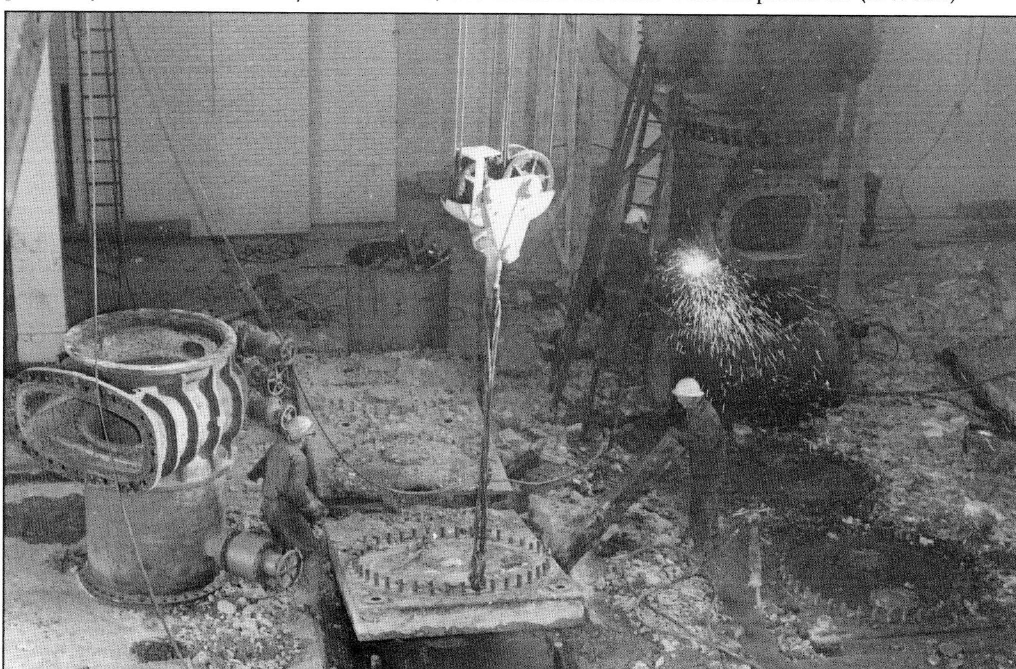

Workers struggle to dislodge the unwieldy base plate of steam-powered Engine 13 on October 13, 1959, while another worker, to their right, uses an acetelyne torch to cut up the discharge water main into smaller, more manageable pieces. (DWSD.)

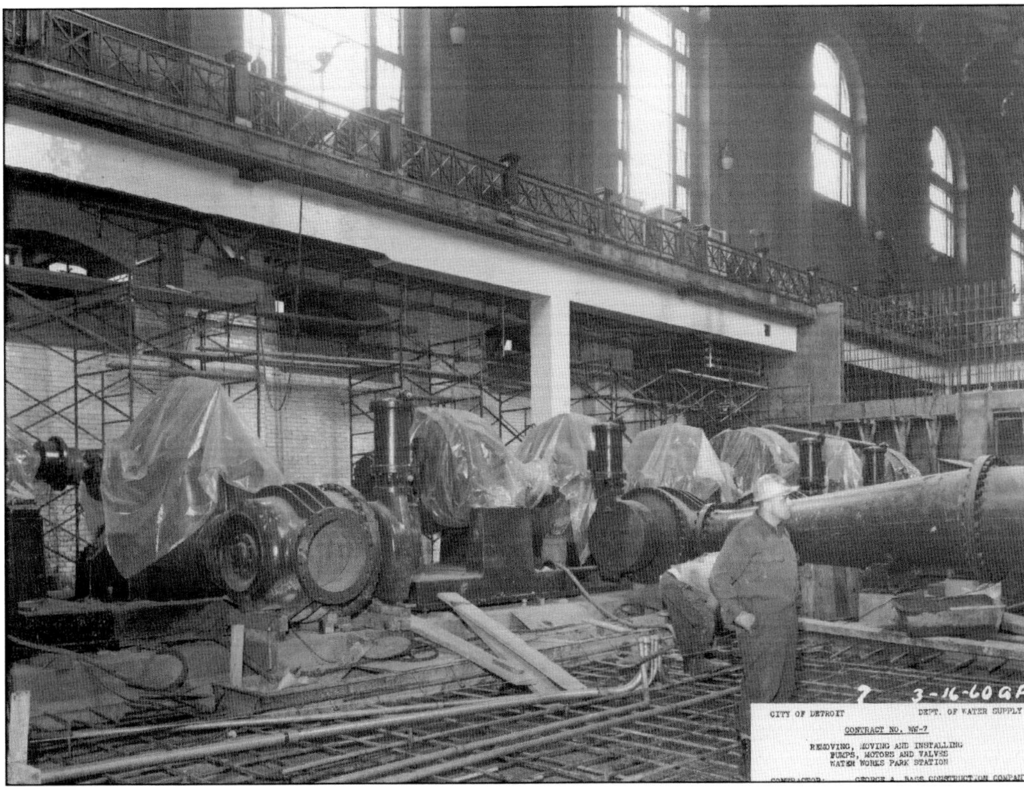

Workers install electric pumps in the High Lift Building on March 16, 1960. Most of them still covered in their factory plastic wrappings, the new pumps look puny and out of place within the cavernous space recently vacated by six steam-driven giants. (DWSD.)

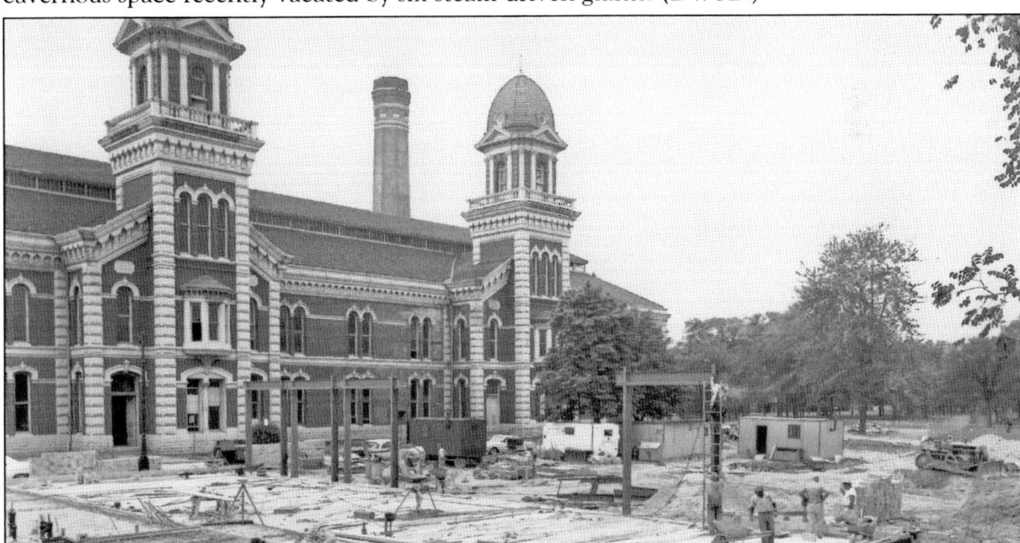

Its interior having already been emptied of the steel giants that made it the beating heart of Detroit's water distribution system, the venerated Pumping Station No. 1 awaits the wrecking ball in July 1962. Above the door on the tower to the left is the legend 1900, the year expansion of the station was begun to make room for more steam engines. (DWSD.)

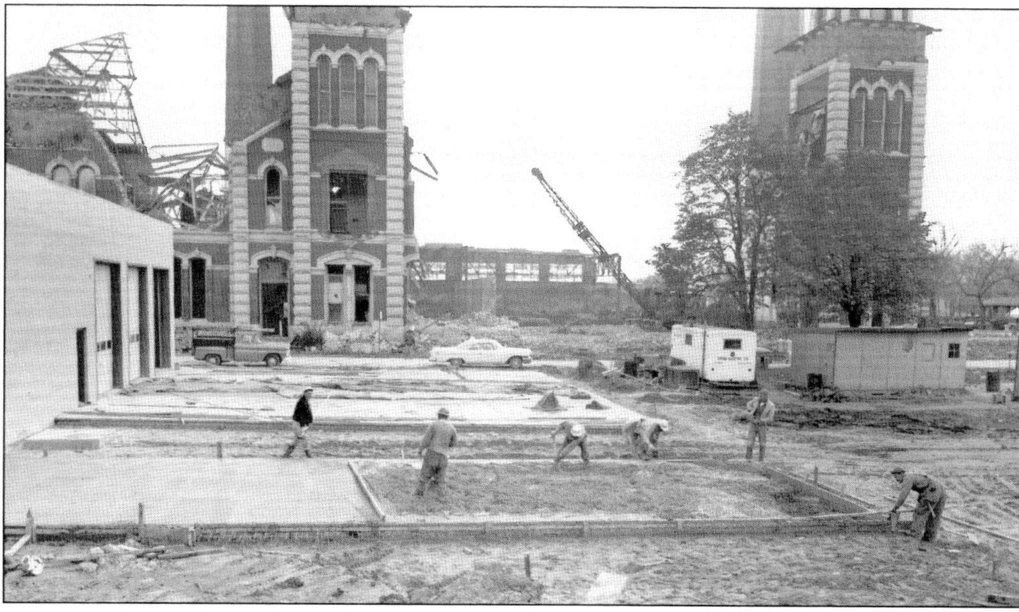

It is out with the old and in with the new as workers lay concrete for a new parking lot in the shadow of Pumping Station No. 1's remains on October 24, 1962. (DWSD.)

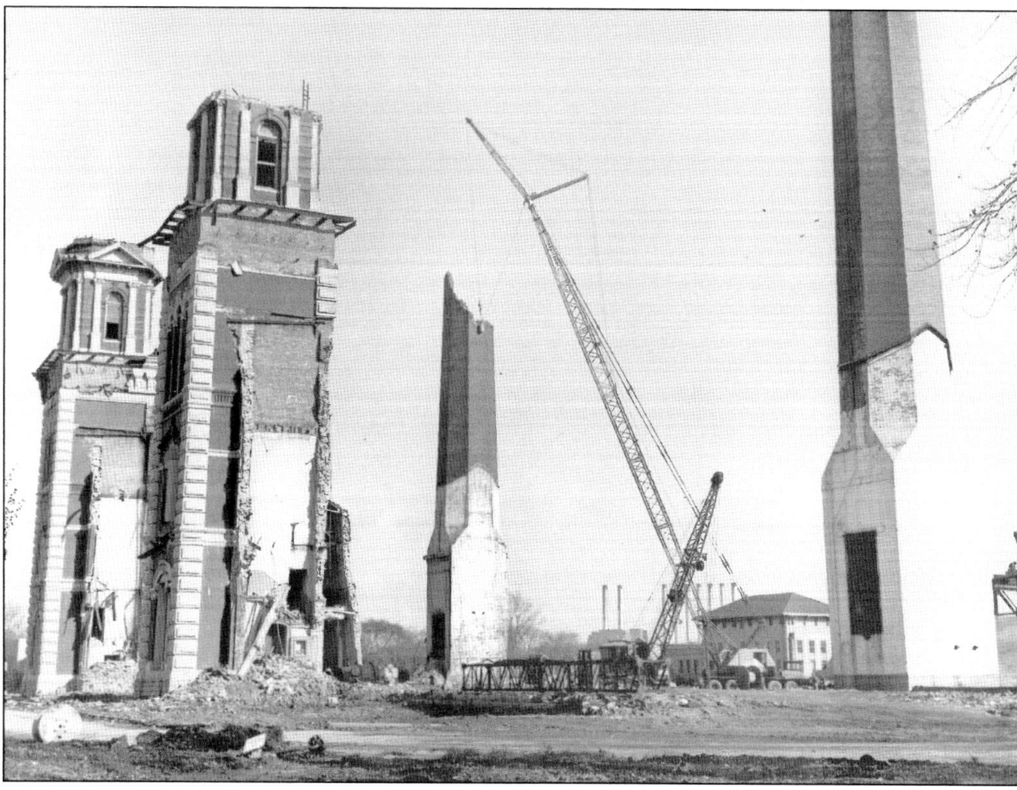

The towers and stacks are all that remain of the original pumping station as they wait for the wrecking ball in 1962 to finish the job. The old chlorination building is shown behind the crane. In the distance are the iconic towers of Detroit Edison's "Seven Sisters" power plant. (DWSD.)

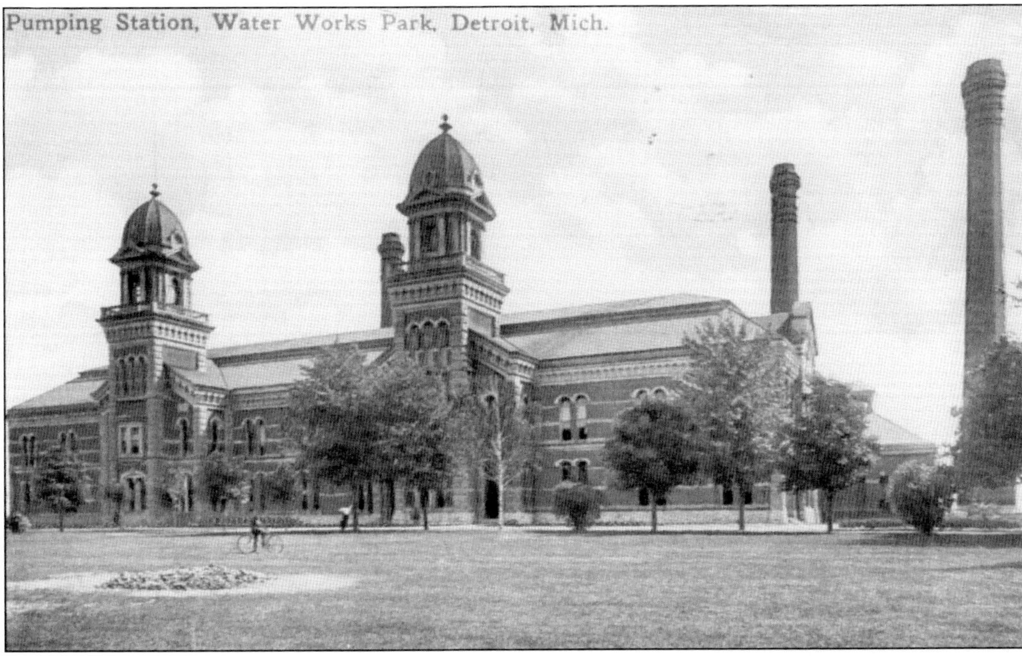

This classic image of Pumping Station No. 1 shows what once was. A relic of Detroit's past, it is gone but not forgotten. (Author.)

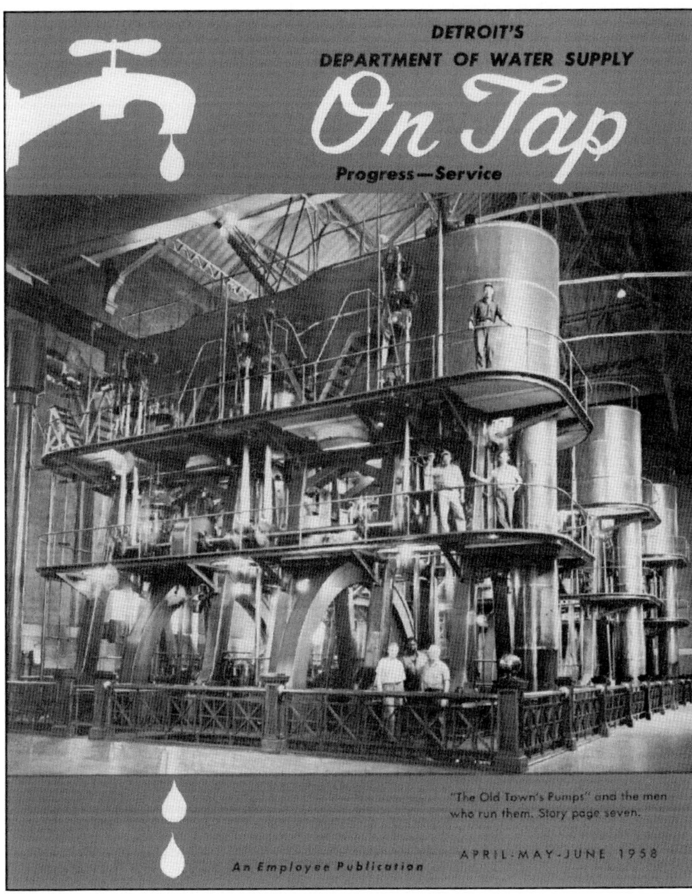

One of the powerful steam engines at Water Works Park graces the cover of the water department's employee newsletter in 1958. Once an essential part of the water distribution system, the steel giants exist now only in old photographs and the fading memories of those old enough to remember seeing them. (Author.)

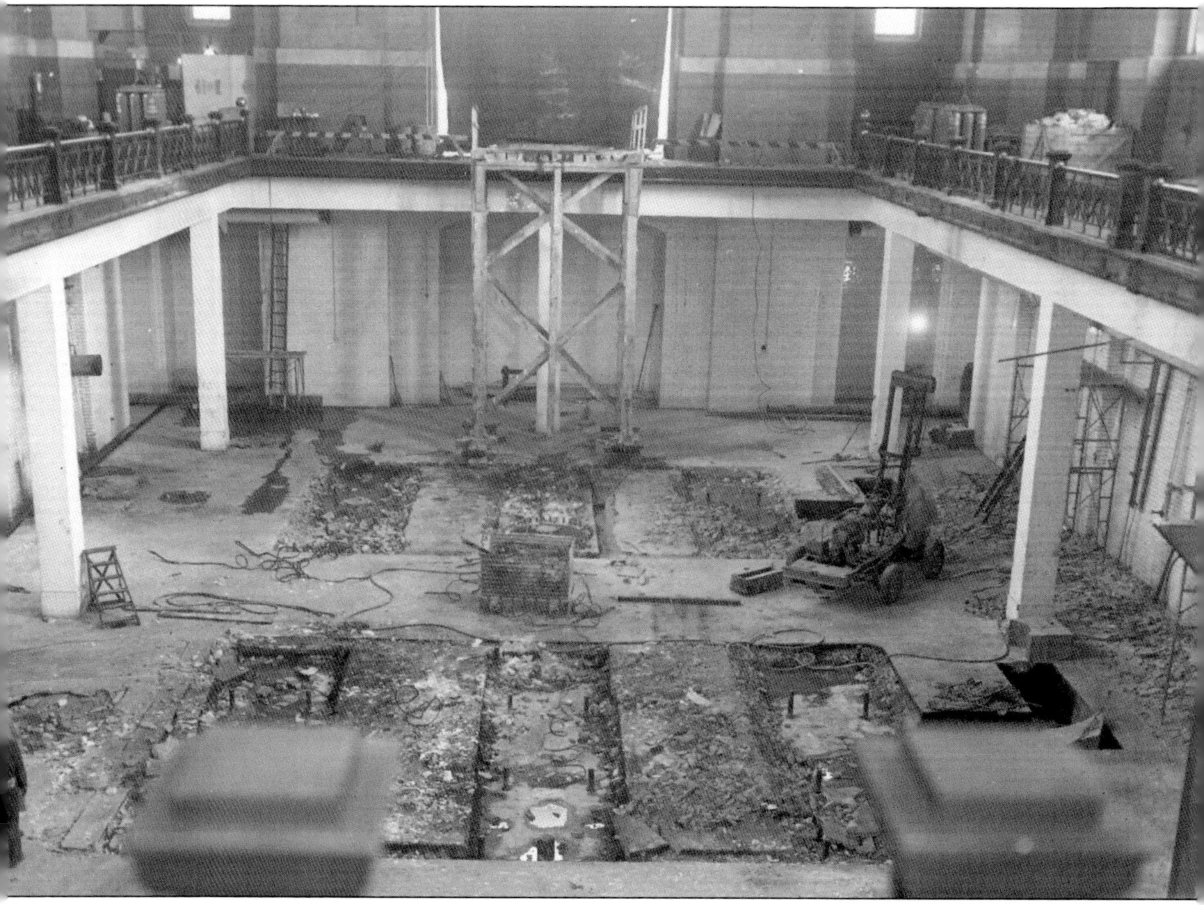

Cleaning up debris in the gutted Pumping Station No. 2 after the removal of the steam-driven pumping engines was a long but necessary process for the switch to electric power. (DWSD.)

Six

National Security

From its beginnings in the 1870s, the primary function of Water Works Park has been to provide pure and wholesome drinking water to those who live and work in Greater Detroit. Although city officials wholeheartedly supported the park's role as a refuge from the dirt and noise of early-20th-century urban life, its operation as a public park remained secondary to its responsibilities as a public utility and guardian of the city's health and welfare. Closures for reasons of national security during World Wars I and II were the ultimate expression of this prioritization.

The gates of Water Works Park were reopened to the public following the end of both wars, but attitudes had dramatically changed by the time North Korean forces crossed the 38th parallel into South Korea in June 1950.

Fears of Communist infiltration were widespread. The House Un-American Activities Committee was busily investigating real and suspected Communists in positions of actual or supposed influence in the US government, and Sen. Joseph McCarthy's witch hunts would soon begin. When the Hurlbut Memorial Gate was closed to the public this time, it was intended to be permanent, with armed guards posted to enforce the lockout.

Four years after the 1953 armistice that brought an end to hostilities in Korea, city officials relented to persistent public demands for the reopening of Water Works Park. Because a majority of the site was now dedicated to water production, however, the public was allowed access to only a few acres along East Jefferson Avenue and a small portion along the Detroit River.

With the exception of the portion of the park along the Detroit River from which hydroplane races can still be viewed, Water Works Park has remained permanently closed to the public since the early 1970s. Plans to open a publically accessible museum and an environmental academy were quashed in the aftermath of the events of September 11, 2001. There are currently no plans to allow members of the general public back on the grounds outside of supervised facility tours.

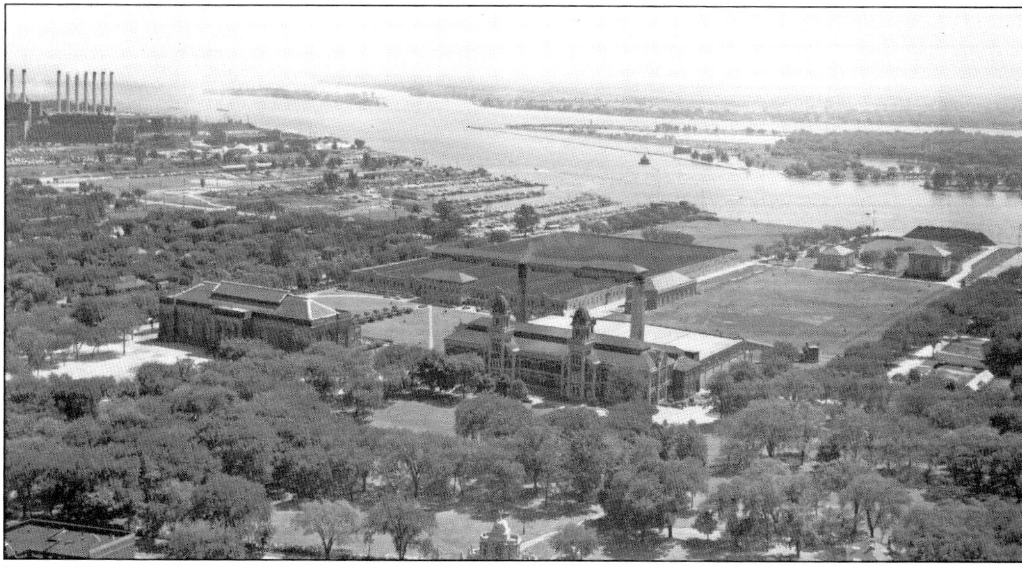

This 1950s aerial view of the park shows the twin towers of the expanded version of the original pumping station to the right. To its left is the High Lift Building, and behind that is Detroit's first water treatment plant (1923–2003). In the distance are the Detroit River, Belle Isle, and Canada. Detroit Edison's iconic "Seven Sisters" power plant is at the far left, with Peche Island behind it in Canadian waters. (DWSD.)

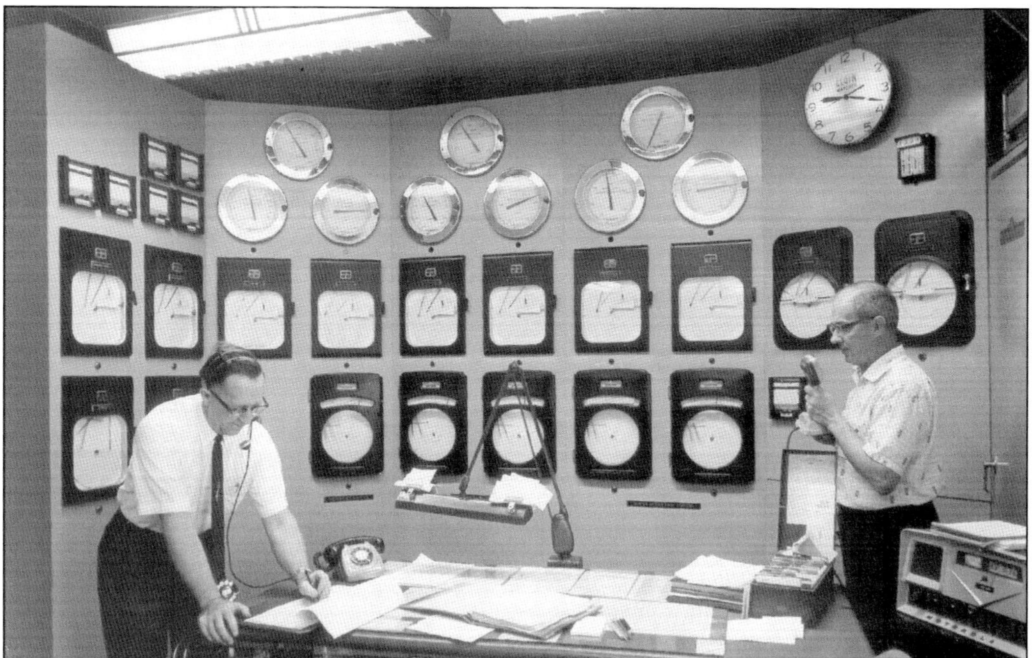

The Systems Control Center for the water distribution system is shown as it appeared in 1964. Introduced in 1956, the SCC (now a Supervisory Control and Data Acquisition, or SCADA, system) has evolved to the point where skilled operators in a central location monitor the entire distribution system, isolate potential trouble spots, and control the pressure in distant parts of the system. (DWSD.)

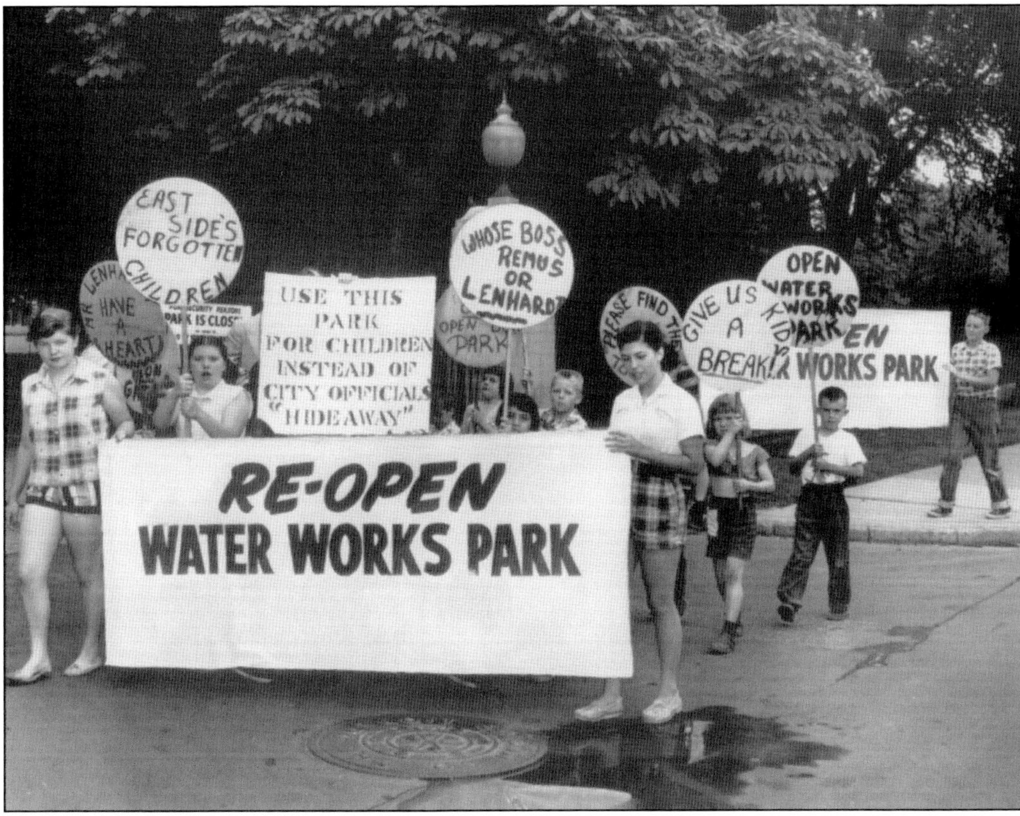

A youthful group of protesters in 1954 demands that the city reopen Water Works Park, closed since the start of the Korean War in 1951. While a small portion of the park was reopened in 1957, it has been permanently closed to the public since the early 1970s. (*Detroit News.*)

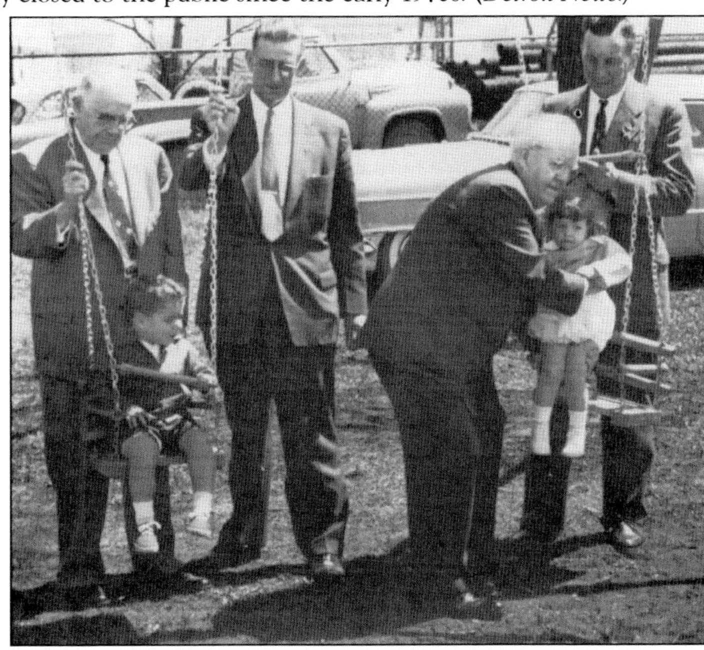

A small item in a 1957 issue of *On Tap*, the water department employee newsletter, announces the reopening of a 7.5-acre portion of Water Works Park to the public on May 1, 1957, in response to popular demand. The playground featured swings and sandboxes for young children, basketball and softball for teenagers, and benches for those less active. (Author.)

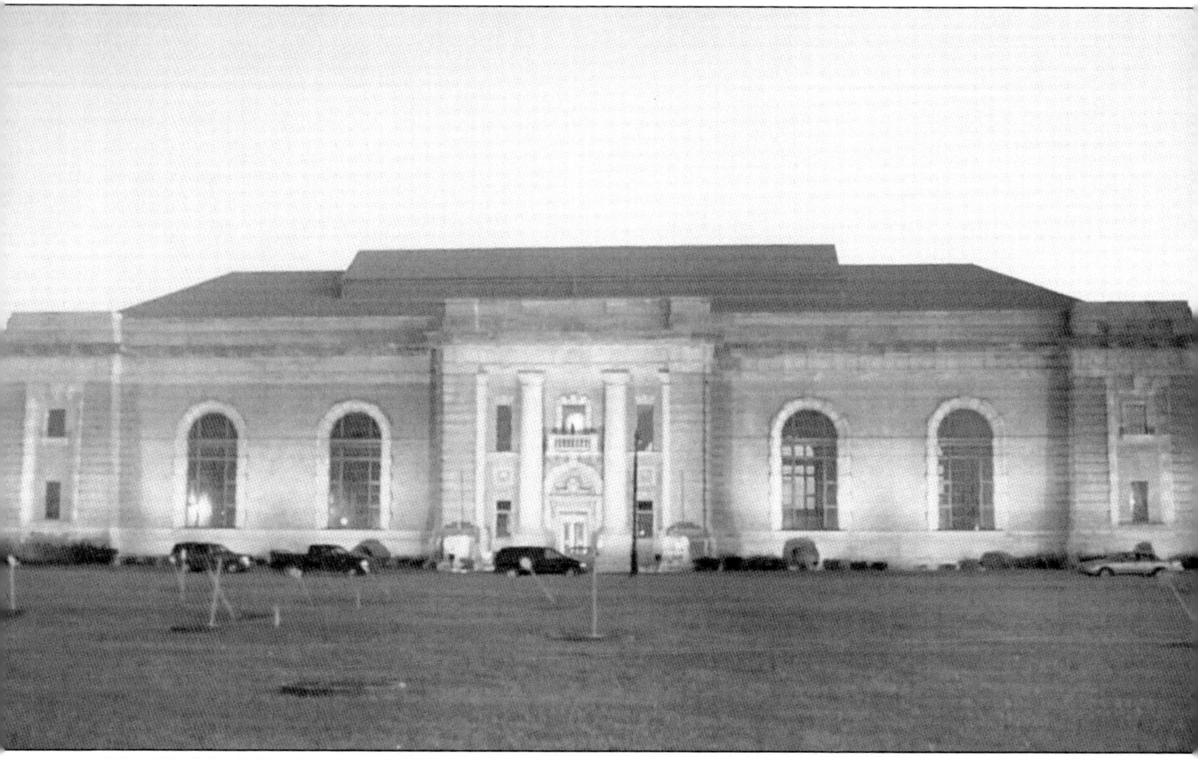

The High Lift Building was designated a Historic Civil Engineering Landmark by the Michigan section of the American Society of Civil Engineers in 2002, the year it is seen here. Originally placed in service in 1914, the building was painstakingly renovated to restore it to its original condition. (DWSD.)

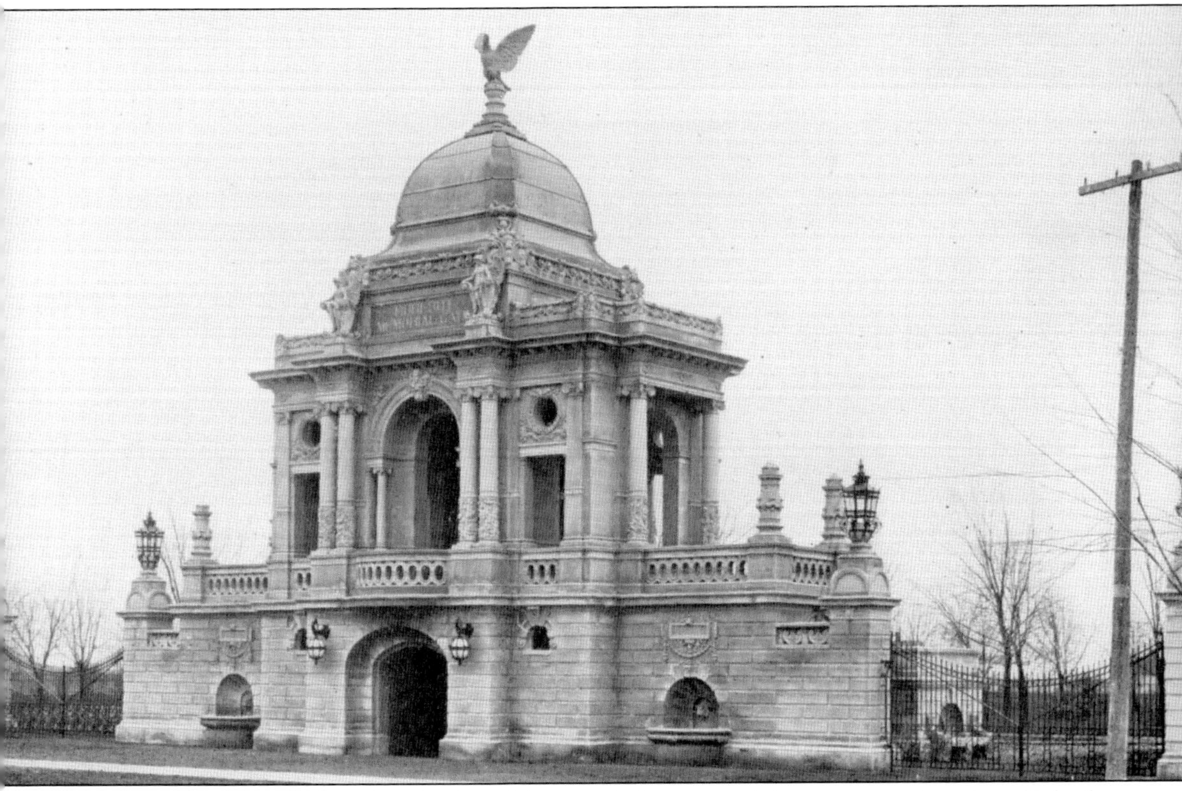

By the end of the 1950s, the Hurlbut Memorial Gate would be one of only a handful of landmarks remaining from the days when the park was one of Detroit's most popular public parks. (Author.)

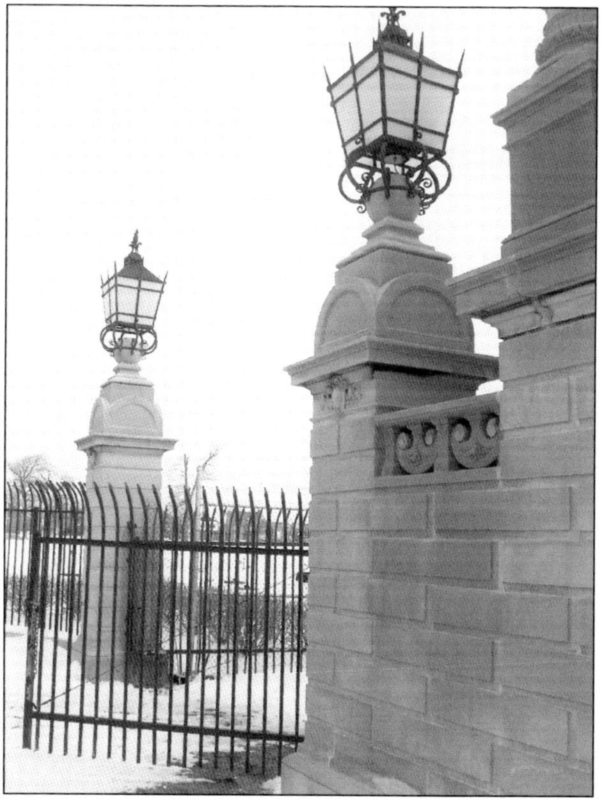

Renovation of the Hurlbut Memorial Gate was undertaken around 2000 to restore decorative items in addition to rejuvenating the structure itself. The decorative lamps fabricated by Carlos Nielbock—a Detroit-based, European-trained, old-school metalsmith—were meticulous re-creations of the original lamps based on vintage images of the gate. Re-fabrication of the originals was necessary as there were no blueprints or plans for the original metal work. The original lamps, fencing, and other decorative items were handcrafted by artisans, often immigrants who learned their craft in the old country. (Left, Carlos A. Nielbock; below, DWSD.)

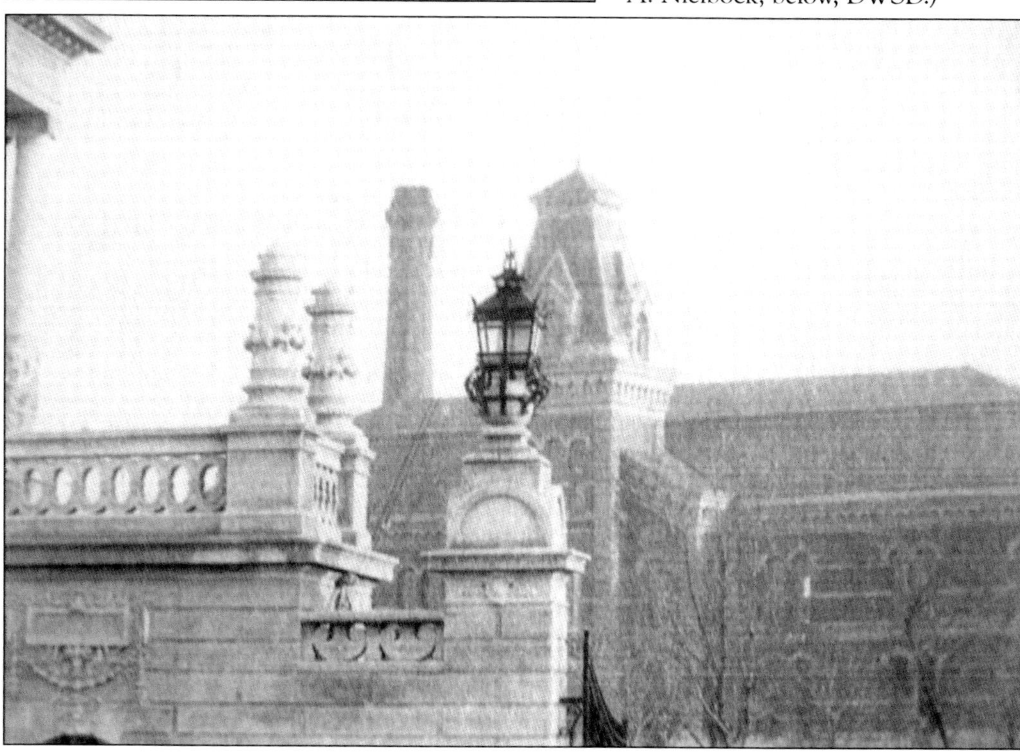

Stephen F. Gorden was director of the Detroit Water and Sewerage Department from 1994 to 2000. During his administration, Detroit started construction of what was then the world's largest water treatment plant to employ ozone as a primary disinfectant. (DWSD.)

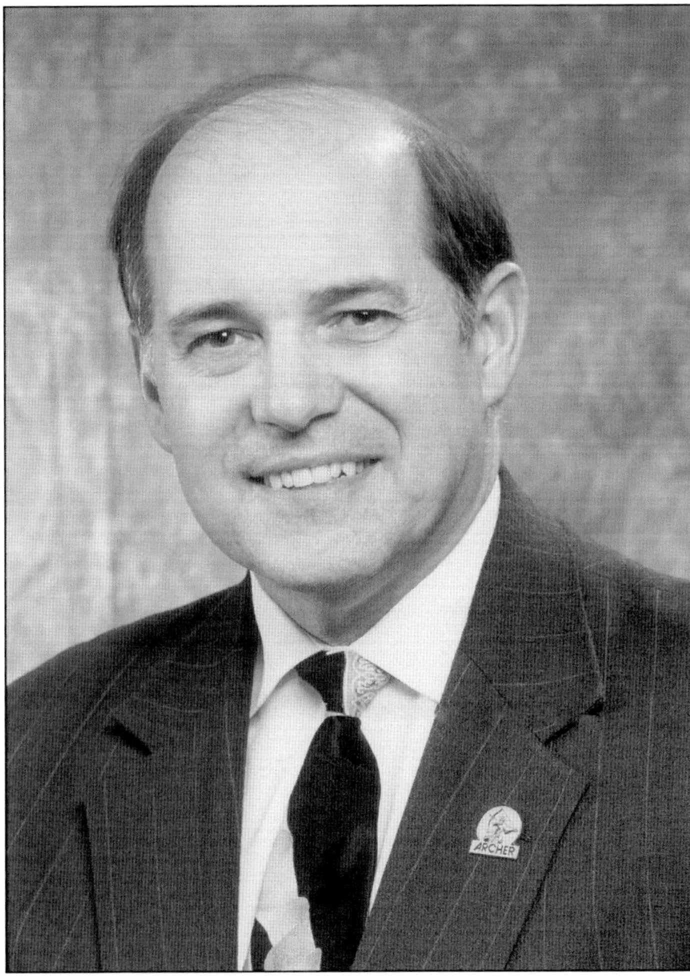

This is a recent image of the administration building in front of the water treatment plant, which began operations in 2003. The Parducci flagpole bases can be seen in front of it. (DWSD.)

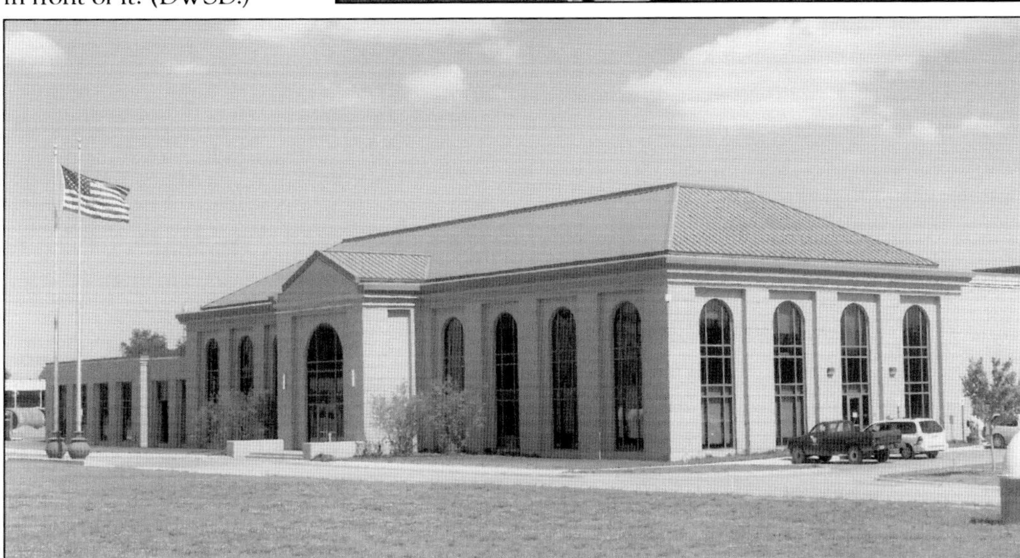

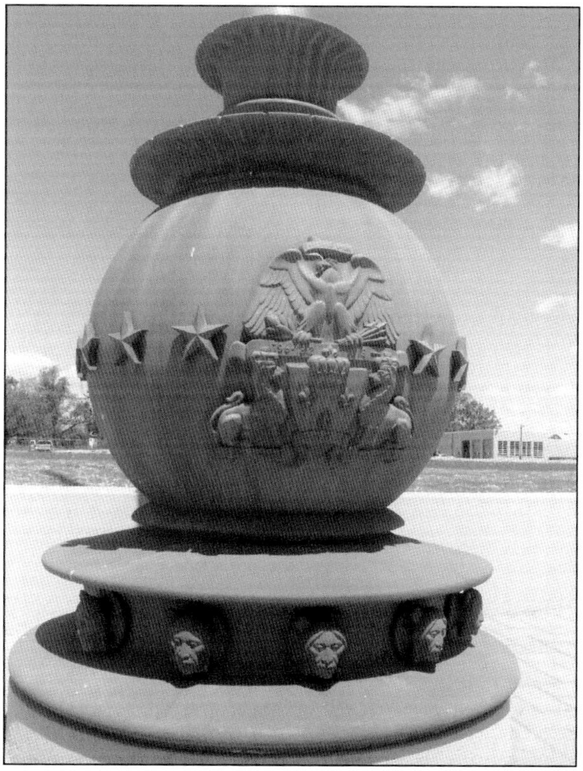

The flagpole bases in front of the new administration building had already acquired a unique history prior to their move to the park. The bases were originally designed by Corrado Parducci for installation in Detroit's Kennedy Square in the 1960s, and they depict Detroit's historic character. The fleur-de-lis on the shield represents the city's French occupation (1701–1760), while the lion represents the periods of British administration (1760–1796, 1813). The eagle, of course, represents the United States, and the 15 stars around the base represent the 15 states at the time Detroit was claimed from the British by Col. John Francis Hamtramck in 1796. The faces of Native Americans around the bottom represents the five Indian nations that held title to the land at the time of the Europeans' arrival. (Both, DWSD.)

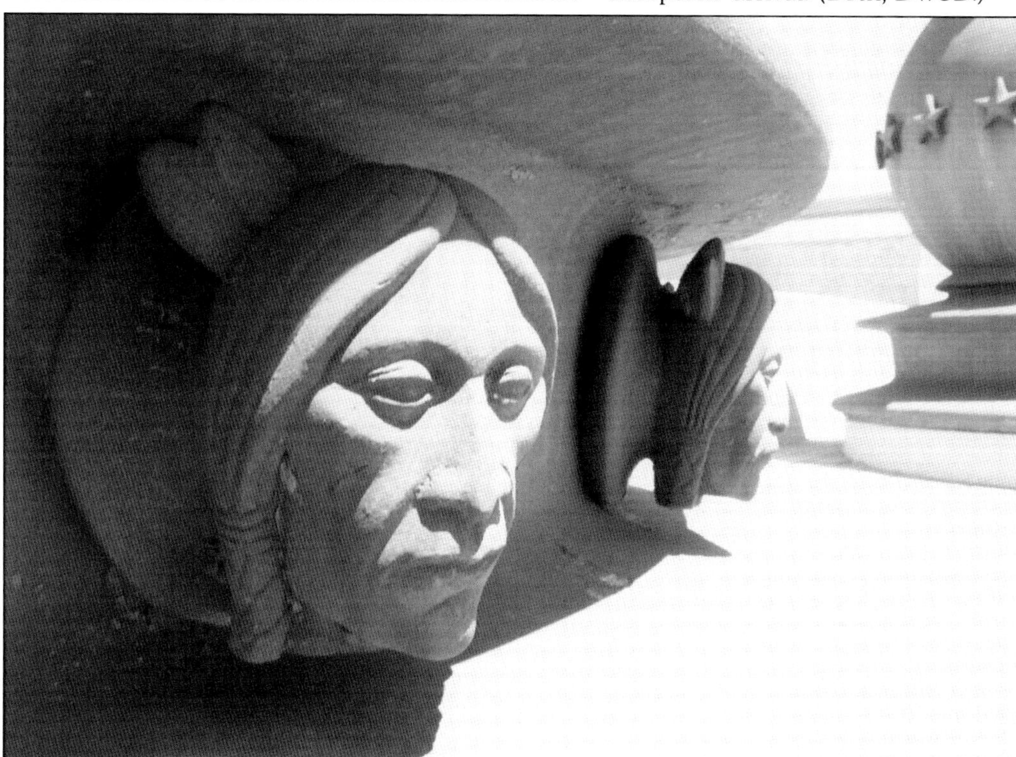

Seven

Souvenirs and Mementos

To quote William Davies King, author of *Collections of Nothing*, "Nearly everyone collects something, even those who don't think of themselves as collectors." The sheer variety of things humans collect is almost limitless: coins, cars, stamps, spoons, postcards, ceramics, comic books, art prints, beer steins, bottle caps, cigar ribbons, and the list goes on and on. People use them to commemorate significant events and visits to special places. Many collect things as a way of revisiting their childhoods. Staying true to their Colonial roots, Americans frequently collect things inspired by their British cousins across the pond.

The earliest-known English commemoratives date from Charles II's restoration as king in 1660. A number of the most common things collected by Americans originated during England's Victorian era. Spoons were one such popular collectible.

Sterling silver commemorative spoons first appeared in 1851 in conjunction with the Crystal Palace Exhibition in London. Marketers created a market and then exploited the demand for souvenir spoons, which travelers purchased as mementos of their visits to noteworthy destinations. The American "spoon craze" peaked during a 30-year period beginning around 1890.

Evolving from illustrations of places and events made on lithographs and woodcuts starting around 1840, collecting postcards is another popular hobby that began around this time. The backs of early picture postcards were exclusively reserved for postage and the address of the recipient. If the sender wished to include a personal greeting, he had to make do with a small area on the front of the card. The "divided back" postcards most of people are familiar with today first appeared in the United States in March 1907.

The earliest souvenir collectibles specific to Water Works Park date from the 1890s, at which time the park was coming into its own as a public park and noteworthy attraction. Souvenirs and mementos collected by visitors to the park included commemorative silver spoons, postcards, glass paperweights, and a variety of ceramic items, including drinking cups, vases, tiny shoes, and commemorative plates.

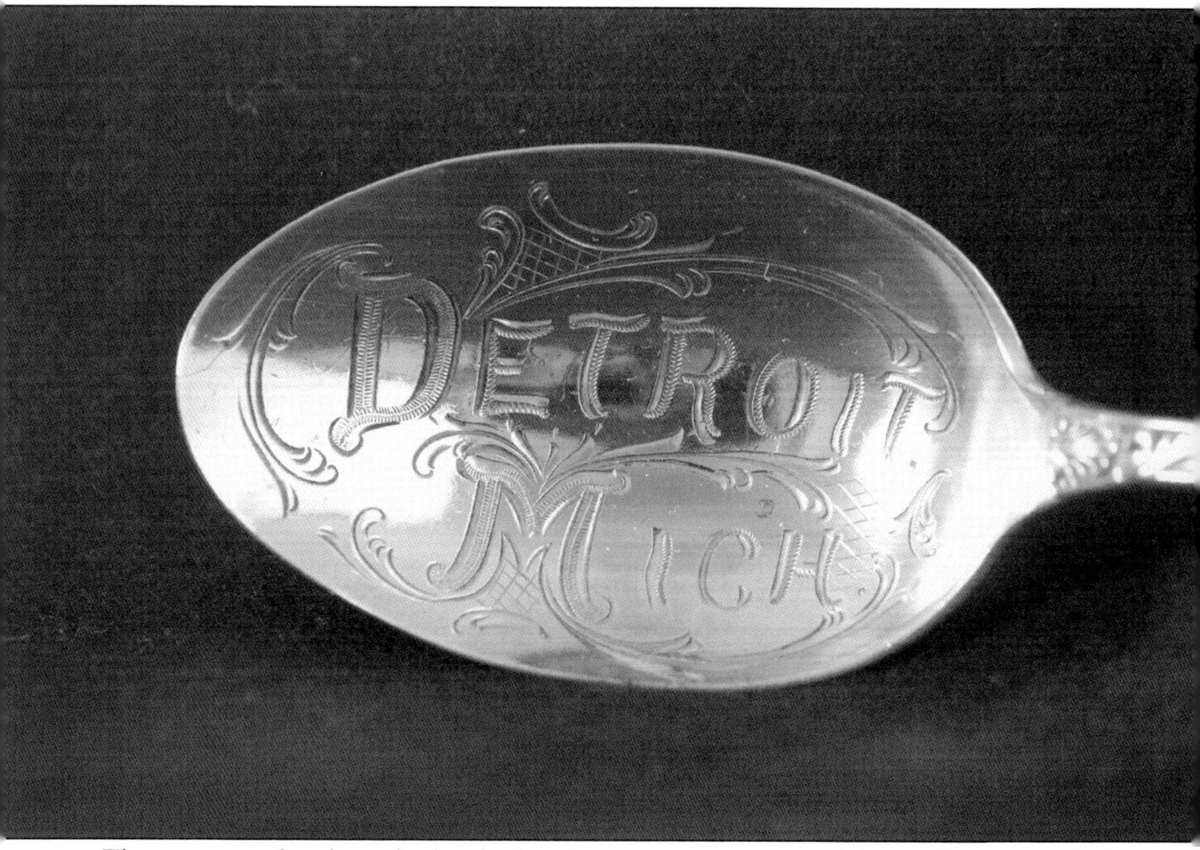

This intricate detail on the bowl of a silver souvenir spoon celebrates Detroit in 1908. (Paul Gomez.)

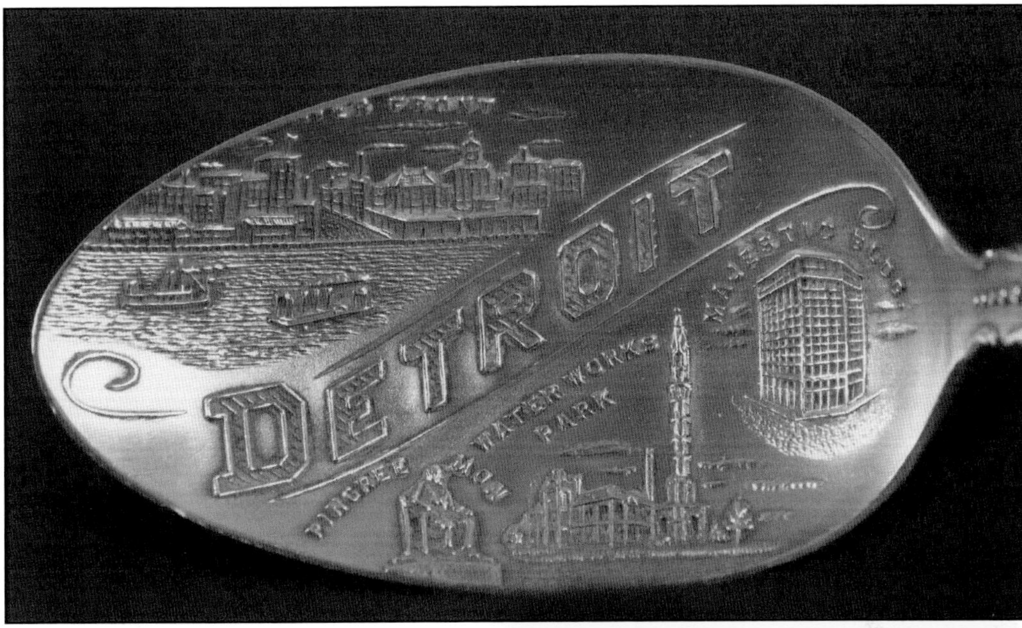

This 1910 sterling silver souvenir spoon depicts noteworthy Detroit attractions, including the Detroit riverfront, the monument to Detroit mayor and Michigan governor Hazen S. Pingree in Grand Circus Park, the pumping station and standpipe at Water Works Park, and the 14-story Majestic Building, Detroit's second skyscraper. (Paul Gomez.)

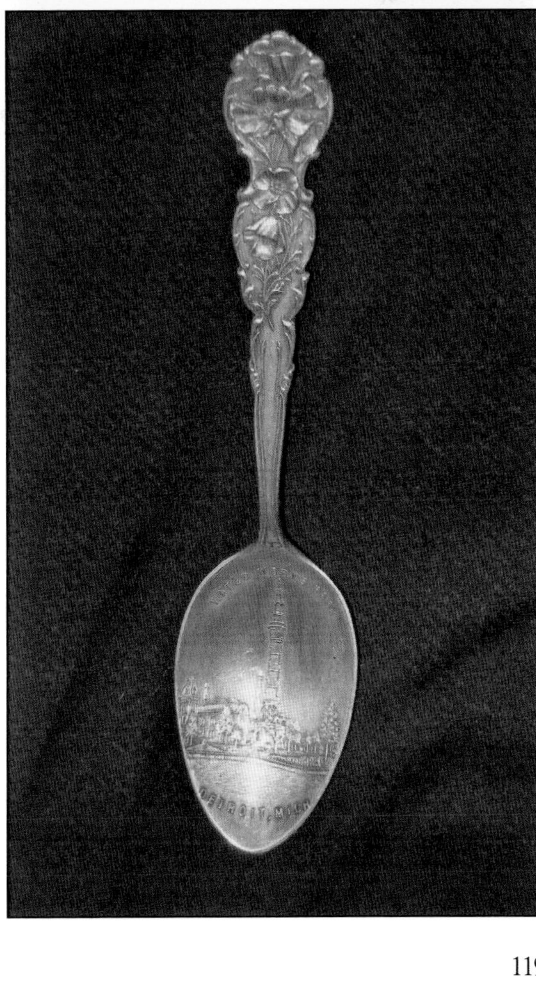

The lake, standpipe, and pumping station at Water Works Park are depicted in sterling silver on this souvenir spoon from the 1890s. Among the trees are some of the Twelve Apostles, French pear trees planted by Detroit's original French settlers in the early 1700s. (Paul Gomez.)

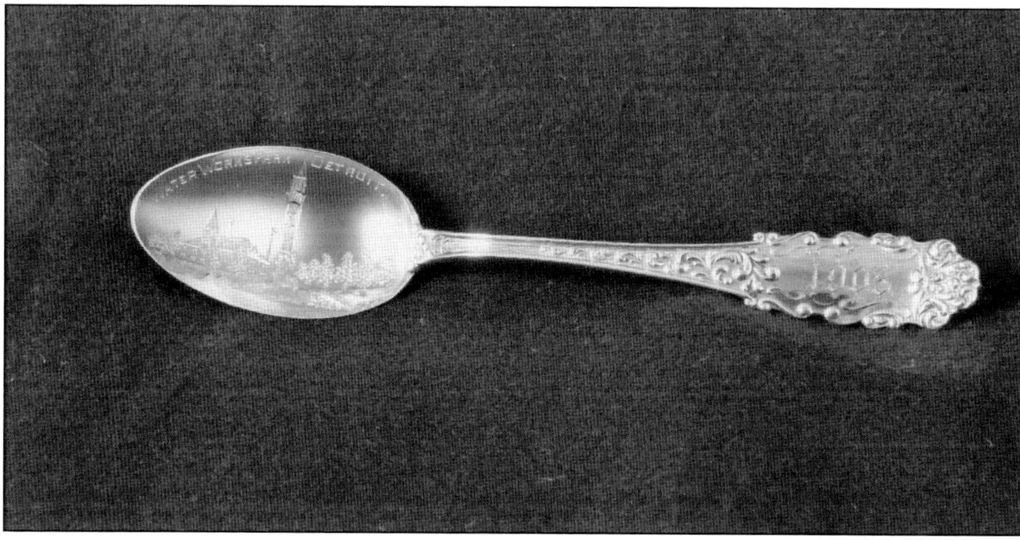

This sterling silver spoon also features the lake, pumping station, and standpipe. Though it is dated 1903, the configuration of the pumping station is pre-1900. (Paul Gomez.)

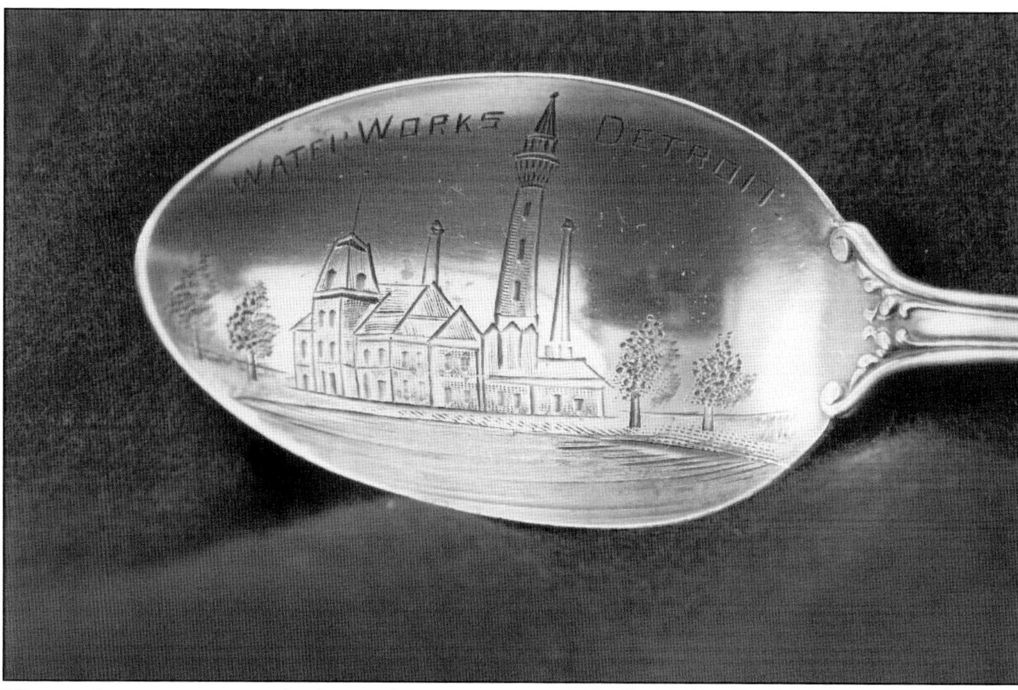

The single entry tower on the front of the pumping station dates this souvenir spoon to the 1890s. The artist, however, has inaccurately depicted the standpipe as being a part of the pumping station. The standpipe was connected to the pumping station by underground pipes but physically separated from it aboveground. (Paul Gomez.)

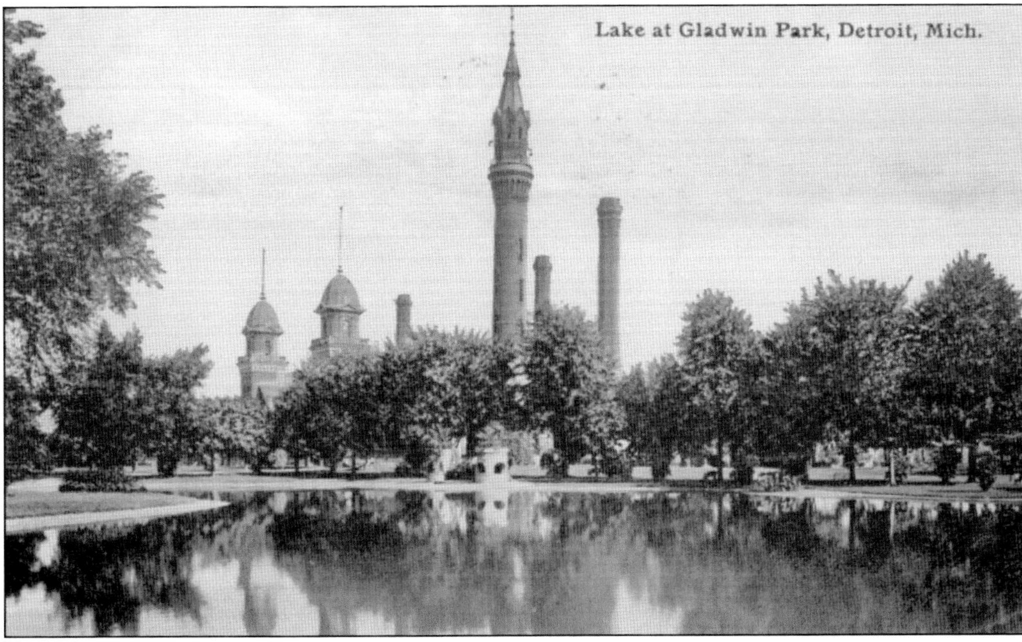

These two postcards of the lake and pumping station, from around 1912, show night and day at the park. These hand-tinted images were among the many postcards popular with tourists as a way to document and share the memories of their visits to the park. Picture postcards have been popular in the United States since about 1900. Official US Post Office Department records for the fiscal year ending June 30, 1908, indicate that nearly 678 million postcards were mailed during that year. That number is especially significant considering the total population of the country at the time was less than 100 million. (Both, author.)

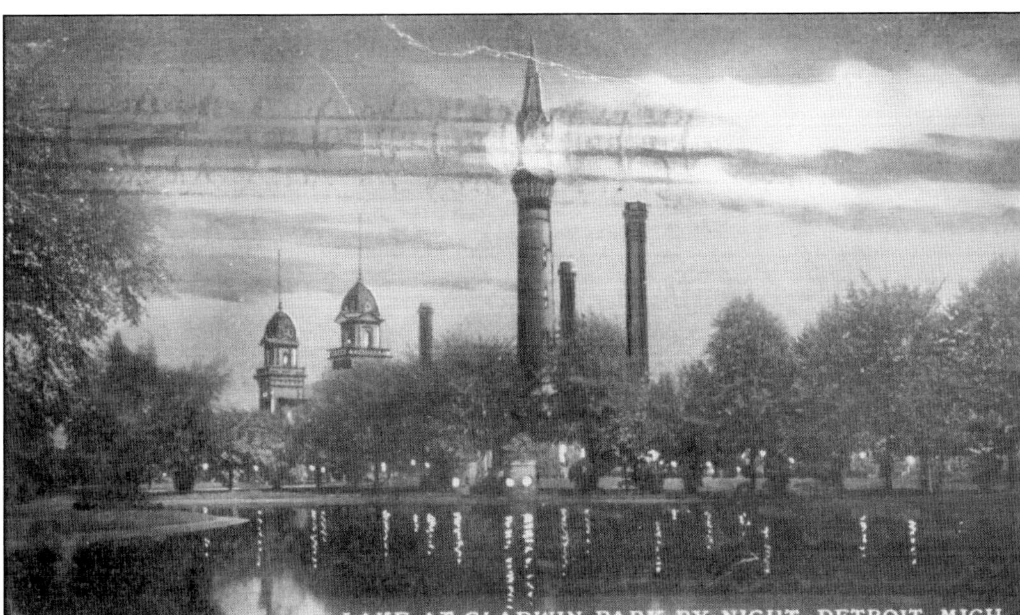

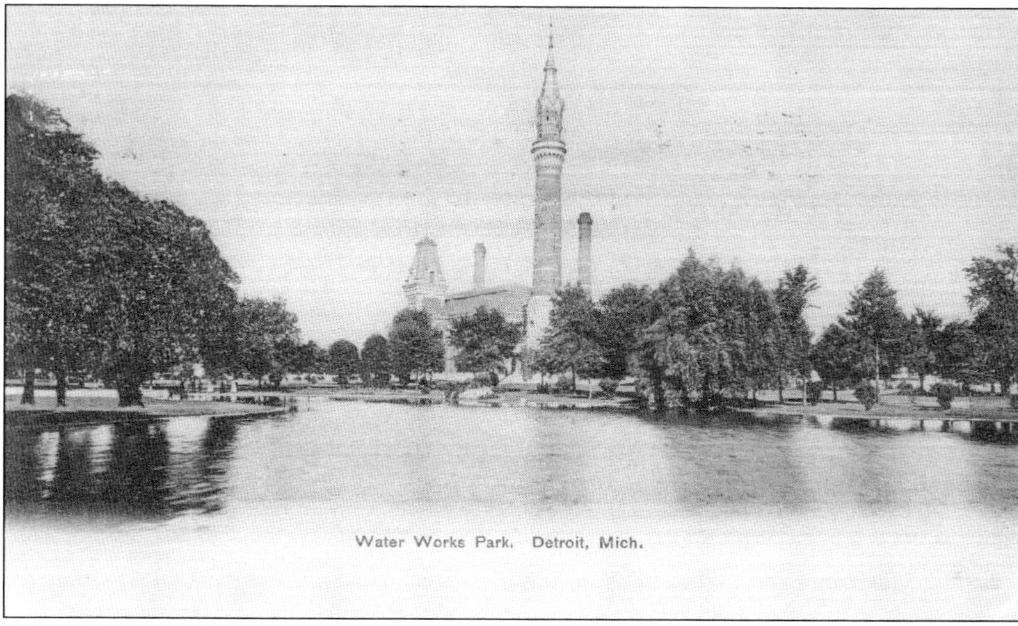

This early postcard from the 1890s of the standpipe and pumping station on the lake is a rare study in black and white. Most postcards from around the turn of the century were hand-tinted variations of the original black-and-white images. (Author.)

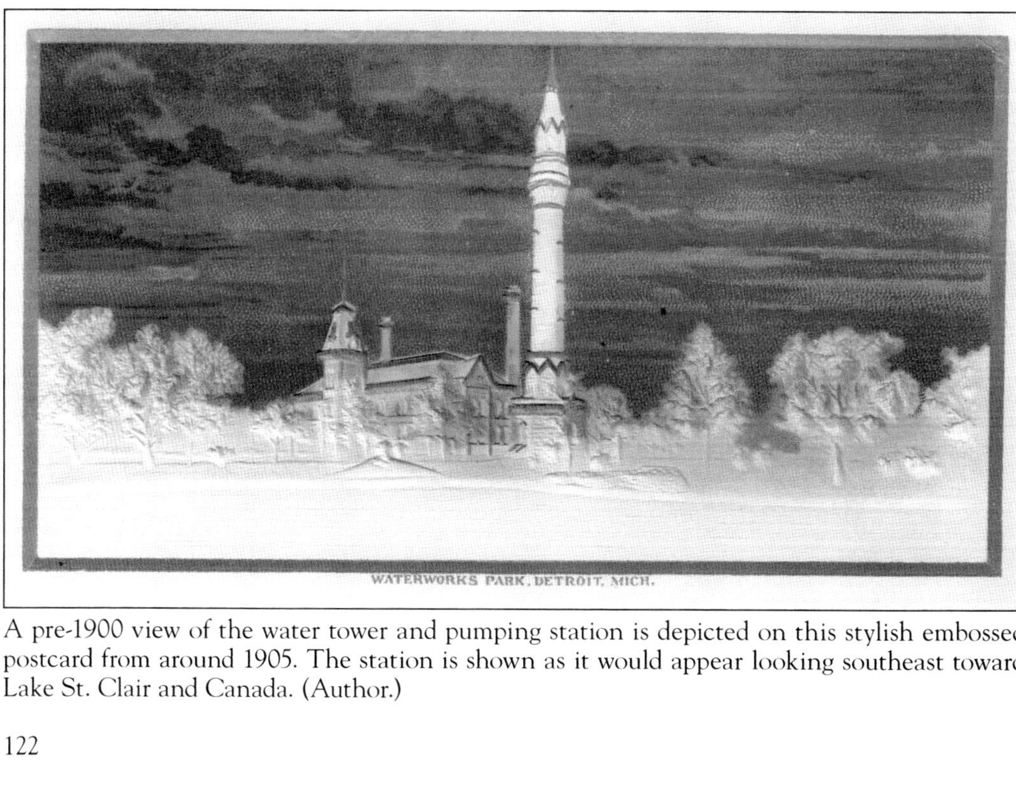

A pre-1900 view of the water tower and pumping station is depicted on this stylish embossed postcard from around 1905. The station is shown as it would appear looking southeast toward Lake St. Clair and Canada. (Author.)

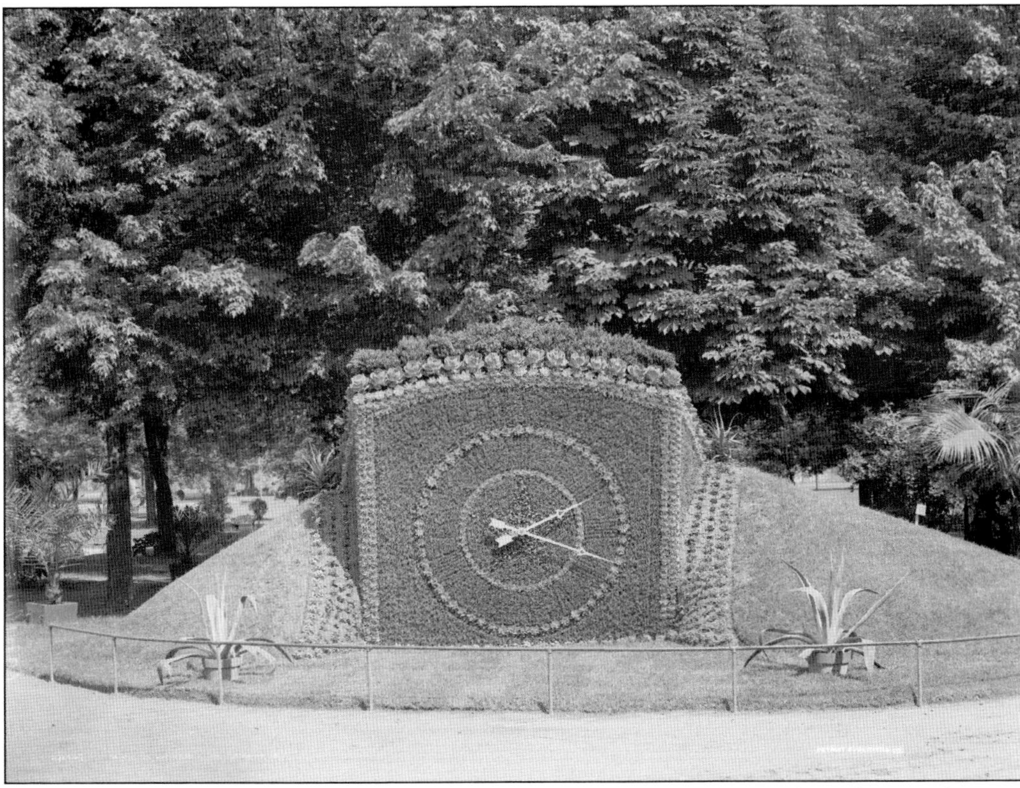

The comparison of these two images of the floral clock face helps explain the popularity of hand-tinted souvenir postcards in the beginning of the 20th century. The star and clock-face numerals, as well as the "Water Power" legend, are indistinguishable from the surrounding floral background in the black-and-white image above. But the colorized view of the clock face on the postcard below is much more compelling, despite the cartoon-like appearance of the people on the walking path. (Above, DPC/LOC; below, author.)

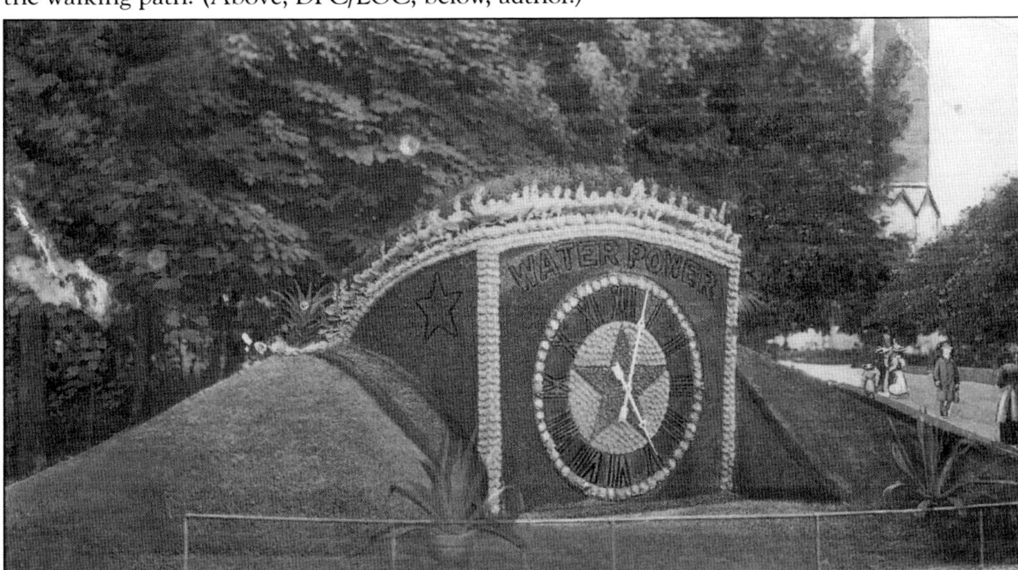

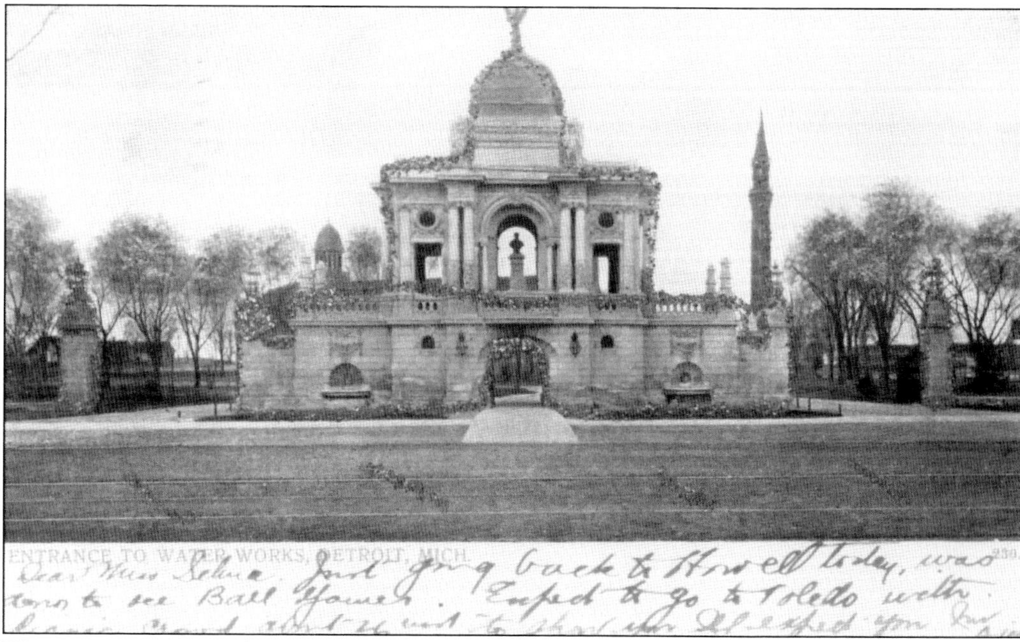

This souvenir postcard of Hurlbut Memorial Gate appears to have been embellished with the decorative application of mucilage and glitter by a young sender. The writing on the front of this card most likely dates it to the 1890s. (Author.)

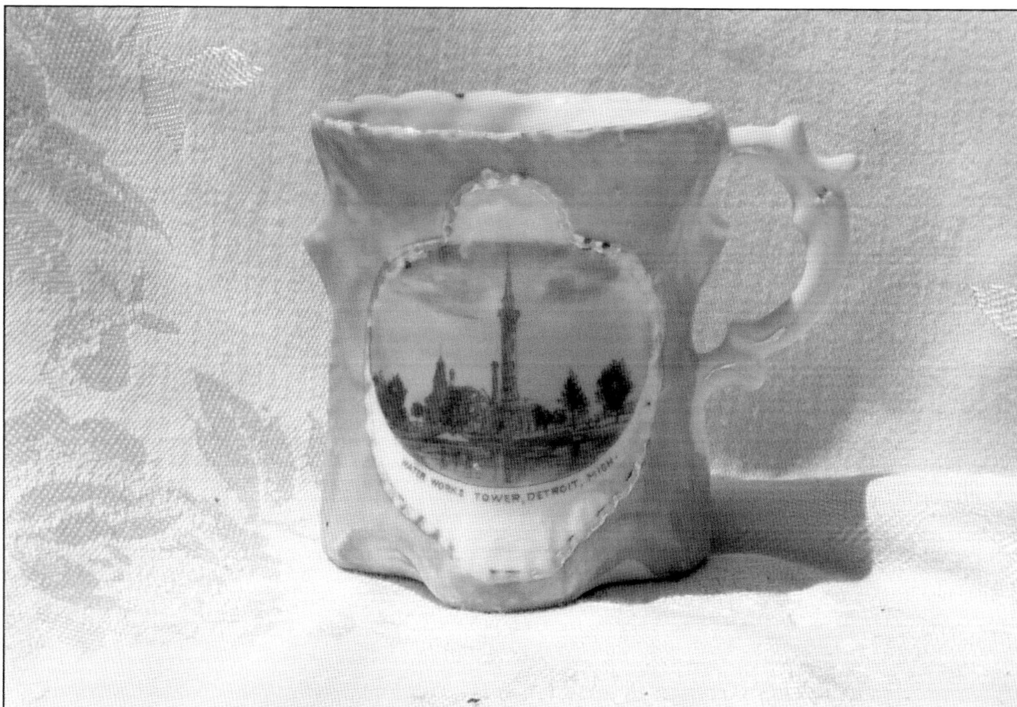

This souvenir ceramic vase, probably dating to the 1880s, shows the lake, pumping station, and standpipe. Though its origin is unspecified, the vase is likely of German origin, as were many of the small ceramic objects produced for the tourist trade at the park. (Author.)

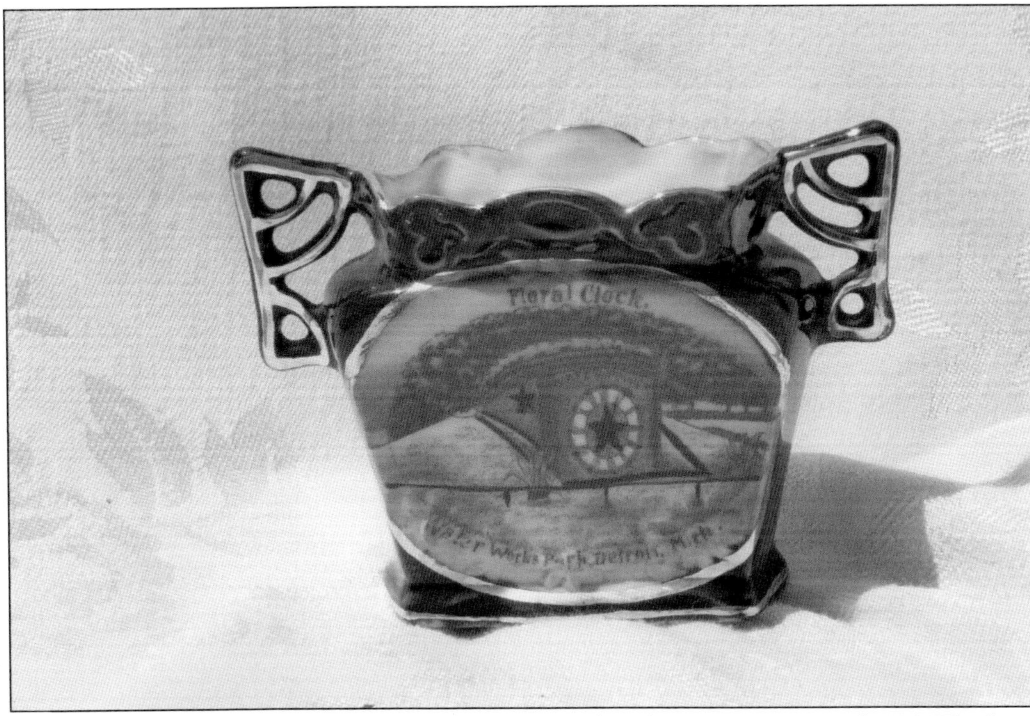

The famous floral clock appears on this flower vase produced in Germany around 1920. The floral clock was a popular attraction at the park from 1893 until 1934, when it was dismantled due to poor time-keeping and high maintenance costs. (Author.)

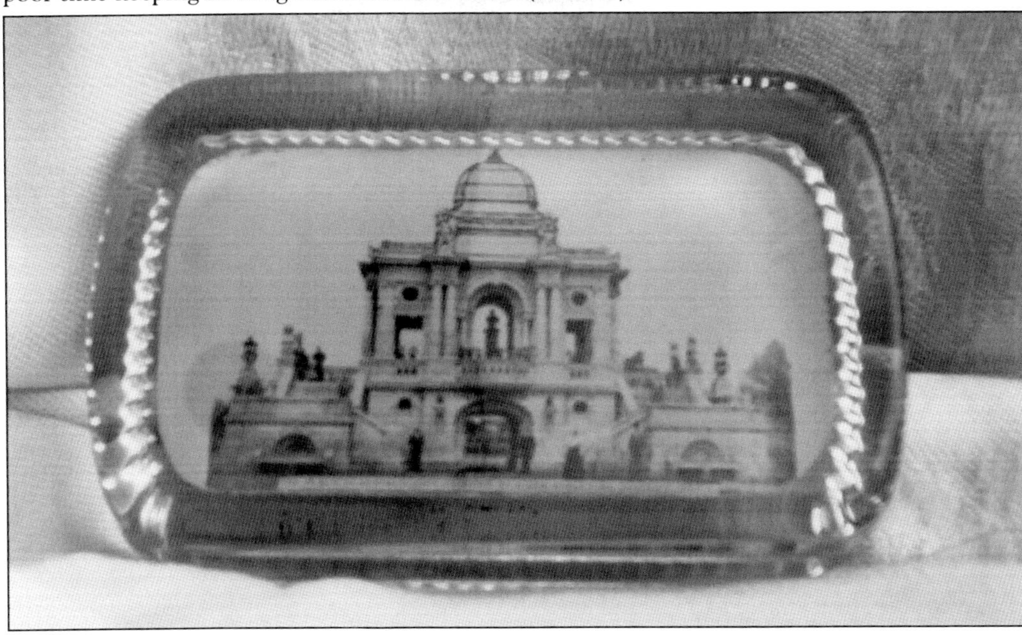

A souvenir glass paperweight contains a fading and rare view of the Hurlbut Memorial Gate, seen in a northward view from inside Water Works Park. The nearly illegible text at the bottom of the monument reads, "The Hurlbut Memorial Gate Entrance to Water Works Park Jefferson Avenue." (Author.)

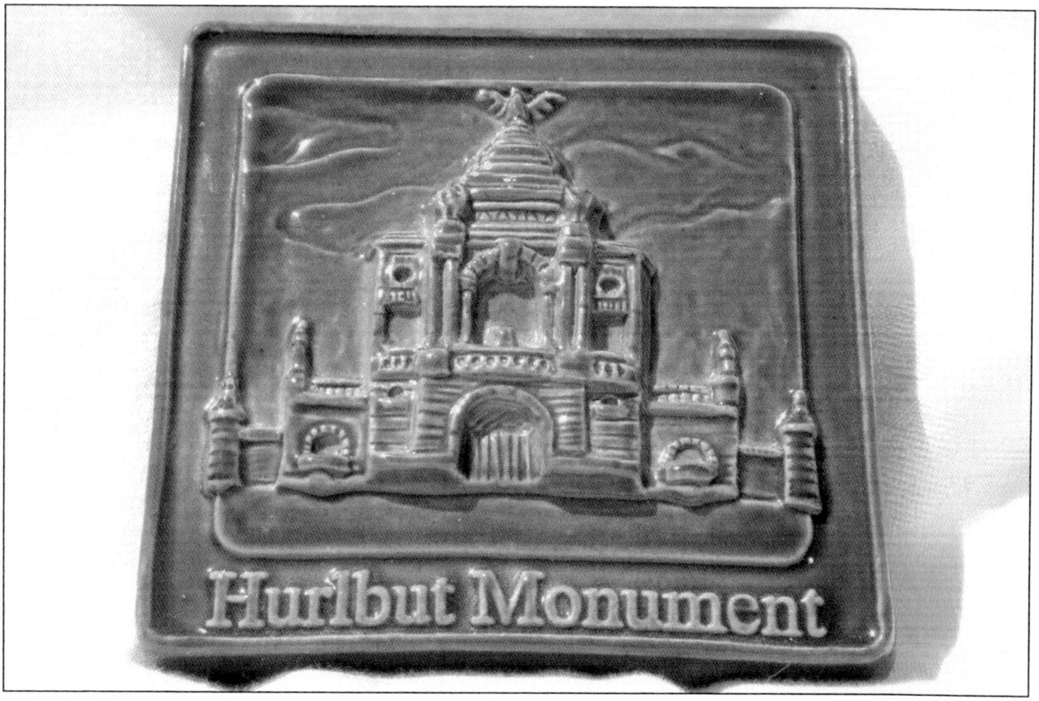

The Hurlbut Memorial Gate and High Lift Building are depicted in commemorative ceramic tiles created by Pewabic Pottery. The tiles were commissioned in 2000 by Hazen and Sawyer, the principal contractor during the construction of the current water treatment plant, which became operational in 2003. The depiction of the Hurlbut Memorial Gate is an accurate rendering of the monument as it currently appears. Note the absence of the bust of Chauncy Hurlbut, which was stolen by vandals in 1974. Pewabic Pottery was founded in 1903 by Mary Chase Perry and moved across the street from Water Works Park into a Tudor building in 1907. It continues to operate within the facility, which was designated a national historic landmark in 1991. (Above, James Heath; below, author.)

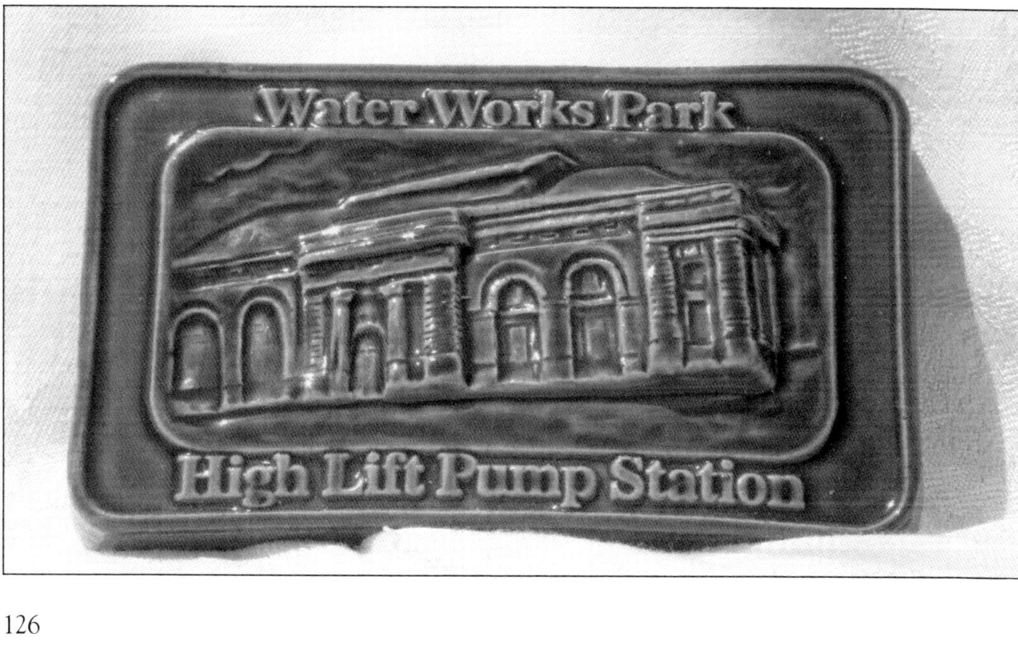

MATERIAL SOURCES

American Water Works Association website. www.awwa.org.
The British Monarchy website. www.royal.gov.uk.
Burton, Clarence Monroe. *History of Wayne County and the City of Detroit, Michigan*. Detroit: S.J. Clarke Publishing Company, 1930.
Engel, Dorothea M. *History of the Board of Water Commissioners*. Detroit: Wayne State University Press, 1937.
Farmer, Silas. *The History of Detroit and Michigan*. Detroit: Silas Farmer & Company, 1884.
Marzejka, Laurie J. "Detroit's Water Works Park a gateway to the past." *The Detroit News*, June 14, 2000.

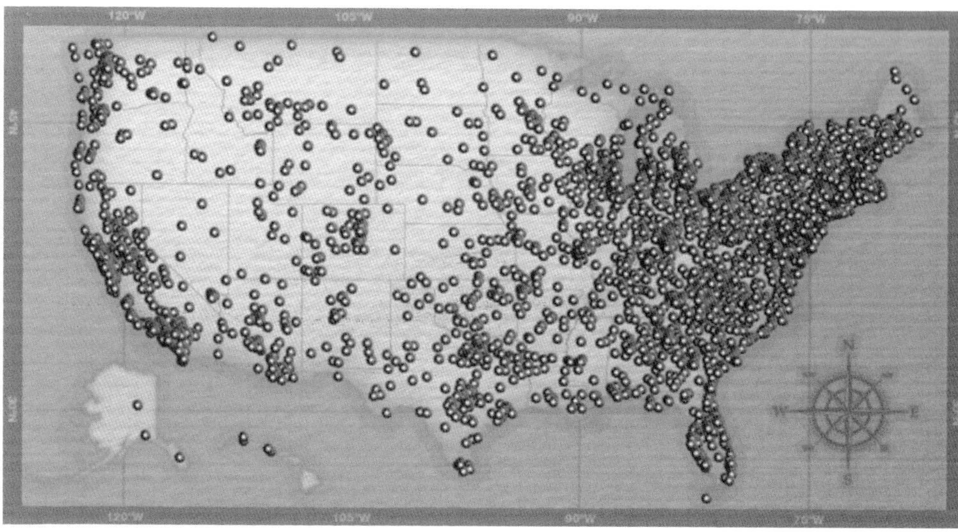